A PLUME BOOK

WHAT ON EARTH?

DR. PETER C. BARNARD

Quentin Wheeler is a professor in sustainability and life sciences at Arizona State University, where he is a Senior Sustainability Scientist in the Global Institute of Sustainability and founding director of the International Institute for Species Exploration (IISE). He was formerly a professor of insect taxonomy at Cornell University, director of the division of environmental biology at the U.S. National Science Foundation, and keeper and head of entomology at London's Natural History Museum. He writes a weekly column, "New to Nature," on species discoveries for *The Observer* newspaper in London. His previous books include *Fungus-Insect Relationships: Perspectives in Ecology and Evolution*; *Extinction and Phylogeny*; *Species Concepts and Phylogenetic Theory: A Debate*; *Letters to Linnaeus*; and *The New Taxonomy*. He lives in Tempe, Arizona, with his wife, Marie, and Chesapeake Bay Retriever, Maddie.

KEITH CRNIC

Sara Pennak holds a master's degree in public administration from the University of Washington and has specialized in public policy analysis, program implementation, research design, and data management during stints at the University of Washington, Penn State University, and Arizona State University. In addition to managing the IISE's popular *State of Observed Species* (SOS) reports and its Top 10 New Species list, she periodically teaches an upper-division course in research ethics for the behavioral sciences called "Scandals, Lies and Torture." Sara cohabitates in Mesa, Arizona's desert uplands with two big dogs; a bad cat; a really good husband, Keith; and an occasional scorpion or two (plus javelinas, snakes, bobcats, a gazillion quail, lizards, coyotes, and a really large roadrunner—all of which stay *outside*).

WHAT ON

EARTH?

100 OF OUR PLANET'S MOST AMAZING NEW SPECIES

Quentin Wheeler
and Sara Pennak

A PLUME BOOK

PLUME
Published by the Penguin Group
Penguin Group (USA) Inc., 375 Hudson Street
New York, New York 10014, USA

USA | Canada | UK | Ireland | Australia | New Zealand | India | South Africa | China
Penguin Books Ltd, Registered Offices: 80 Strand, London WC2R 0RL, England
For more information about the Penguin Group visit penguin.com

First published by Plume, a member of Penguin Group (USA) Inc., 2013

CIP data is available.
ISBN 978-0-452-29814-9

Printed in the United States of America
10 9 8 7 6 5 4 3 2 1

Set in Diverda Serif Com with DIN Next
Map illustrations and book design by Daniel Lagin

We dedicate this book to the scientific laboratories, natural history museums, botanical gardens, and aquaria around the world where taxonomists tirelessly pursue the vision of Linnaeus to inventory and preserve our world's microbes, flora, and fauna.

This book is also dedicated to baby Jack, who timed his arrival right after deadline, and to his adorable big brother, Tim. The world is waiting for you guys—explore!

Contents

Introduction xiii

Chapter 1

PRETTY COUNTS: THE PRETTIEST NEW SPECIES 1

Kaiser's Nudibranch 5
Haisa Orchid Bee 7
Kovach's Orchid 9
Diamantina Tarantula 11
Patton's Bright Snake 13
Fried Eggs Worm 15
Psychedelic Frogfish 17
Wilson's Blue-Eyed Cuscus 19
Barbie Pagoda Fungus 21
Exquisite Sea Urchin 23

Chapter 2

STRANGER THAN (SCIENCE) FICTION: THE STRANGEST NEW SPECIES 25

Little Fork Orchid 29
Quechuan Broad-Nosed Bat 31
Gomes's Freshwater Stingray 33
Double-Hooked Anglerfish 35
Dumbo Octopus 37
Big Brain Protist 39
Little Grooves Earthstar 41
Long-Neck Assassin Spider 43
Sahyadri Nose Frog 45
Ausubel's Mighty Claw Lobster 47

Chapter 3

LESS IS MORE, MORE OR LESS: THE SMALLEST NEW SPECIES 49

Roosmalen's Hairy Dwarf Porcupine 53
Vanessa's Bamboo 55
Pernambuco Pygmy-Owl 57
Smallest Crustacean 59
Obese Diatom 61
Heckford's Midget Moth 63
Cyprus Mouse 65
Teensiest Chameleon 67
Child of Cypris Tiny Fish 69
CSIRO's Medusa Jelly 71

Chapter 4

BIG DEALS: THE LARGEST NEW SPECIES 73

Golden V Kelp 77
Big Red Jelly 79
Raptor Fairy Shrimp 81
Solórzano's Velvet Worm 83
Udzungwa Gray-Faced Sengi 85
Huntsman Spider 87
Chan's Mega-Stick 89
Sierra Madre Monitor Lizard 91
Sir Raffles's Showy Flower 93
Idip's Starfish 95

Chapter 5

SOMETHING OLD, SOMETHING NEW: THE OLDEST NEW SPECIES 97

Finney's Bat 101
Aurora Horseshoe Crab 103
Sarmatian Seahorse 105
Sahel Man 107
Levant Octopus 109
Burma Bee 111
Old-Old Mushroom 113
Walking Cactus 115
Half-Shell Turtle 117
Dila's Flower 119

Chapter 6

HELLO, GOOD-BYE: THE MOST ENDANGERED NEW SPECIES

HELLO, GOOD-BYE: THE MOST ENDANGERED NEW SPECIES 121

Tennessee Bottlebrush Crayfish 125
Martha's Pink Iguana 127
Pygmy Three-Toed Sloth 129
Cloudy Suckermouth Armored Catfish 131
Parecis Lizard 133
Isidoro's Chewing Louse 135
Giant Ribbed Clam 137
Matilda's Horned Viper 139
Strydom's Yam 141
Burrunan Dolphin 143

Chapter 7

LETHAL WEAPONS—VENOMS, TOXINS, AND DISEASE: THE DEADLIEST NEW SPECIES

LETHAL WEAPONS—VENOMS, TOXINS, AND DISEASE: THE DEADLIEST NEW SPECIES 145

Leprosy Bacterium 149
Doris Swanson's Poison Dart Frog 151
Corredor's Assassin Bug 153
Zombie Ant Fungus 155
Lilian's Widow Spider 157
Ashe's Cobra 159
Attenborough's Pitcher 161
Andre Menez's Cone Snail 163
Poisonous Predaceous Polyclad 165
King's Deadly Jelly 167

Chapter 8

GOING TO EXTREMES: NEW SPECIES FROM THE MOST EXTREME ENVIRONMENTS 169

Cave Pseudoscorpions 173
Morafka's Desert Tortoise 175
Yellowstone Bacterium 177
Nature Conservancy Diving Beetle 179
Cryptic Forest-Falcon 181
Stephenson's Antarctic Flower 183
Nepalese Autumn Poppy 185
Bare-Faced Bulbul 187
Siau Island Tarsier 189
Heat-Loving Tonguefish 191

Chapter 9

THE HIGHEST FORM OF FLATTERY: THE BEST NEW SPECIES MIMICS 193

Groves's Nudibranch 197
Firefly Flasher 199
Appalachian Tiger Swallowtail 201
Mache Mountains Glass Frog 203
Fish Mime 205
Ngome Dwarf Chameleon 207
Shocking Pink Millipede 209
Oria's Leaf Insect 211
Denise's Pygmy Seahorse 213
Jumpin' Spider Ant-Mimic 215

Chapter 10
WHAT'S IN A NAME?: NEW SPECIES WITH THE BEST NAMES 217
Apparating Moon-Gentian 221
Madonna's Water Bear 223
Bonaire Banded Box Jelly 225
John Cleese's Woolly Lemur 227
Google Ant 229
David Bowie's Spider 231
SpongeBob SquarePants Fungus 233
Clare Hannah's Shrimp 235
Groening's Sand Crab 237
Wonderfully Photogenic 'Pus 239

Conclusion 241

Acknowledgments 243

References and Credits 245

Index 265

Introduction

We don't know for sure how many species there are,
where they can be found or how fast they're disappearing.
It's like having astronomy without knowing where the stars are.

—E. O. WILSON, *TIME* (OCTOBER 13, 1986)

What do you call 2 million slithering, crawling, running, flying, digging, oozing, and swimming species? A good start. Scientists estimate that no fewer than 10 million kinds of living plants and animals remain to be discovered, and if we include single-celled microbes in our estimate, then the number is significantly higher. Many people are surprised to learn of the steady progress being made by species explorers, who document an average of eighteen thousand new species each year.

Quietly, effectively, and mostly out of the public eye, a dedicated army of professional and amateur taxonomists advances our knowledge of the diversity of life forms and their distribution in the biosphere. They carry on one of the most ambitious and longest-running scientific enterprises ever conceived. While formal efforts to classify the kinds of living things have existed from the time of Aristotle in ancient Greece, "modern" taxonomy began at the height of the Enlightenment in the 1750s with the groundbreaking work of Carl von Linné, better known as Linnaeus.

Incorporating the many theoretical and practical advances made since Linnaeus, taxonomy is the branch of biology focused on species exploration. Taxonomy goes far beyond the initial discovery of new species. After a new species is discovered, a carefully crafted hypothesis is written to discuss the unique combination of features that

Wedding portrait of Carl Linnaeus (1739).

MARK A. WILSON (DEPARTMENT OF GEOLOGY, COLLEGE OF WOOSTER)

differentiates this new life form from other existing species. The species is then classified based on its position on the tree of life using what is known as a cladistic analysis. A cladistic analysis determines where the species fits in the grand scheme of all life forms and allows taxonomists to place each species alongside its nearest relatives. Unique names are then provided to the species, so scientists can record and retrieve information about species and groups of species (taxa). All species are given scientific, two-part names (binomials) that are commonly Latin words, although some new names are rooted in languages as diverse as Spanish, Greek, Quechua, and Sanskrit. The official scientific name for a species is not the same as the species' "common name," which most people use when talking about familiar plants and animals. It may be apples and oranges to most folks, but to scientists they're *Malus domestica* and *Citrus sinensis*.

Only when a name and formal description are published and made available to the scientific community does a new species officially exist. This is not the end of species exploration but, rather, the beginning of a long and ever-evolving process. Every time a taxonomist collects a new specimen, finds another related species, or discovers a new anatomical characteristic, there is a chance to further explore the species. Scientists continually test the predictions about each species' characteristics and uniqueness and, over time, modify and improve the accuracy and fullness of the species' description.

So what is a new species? Taxonomists are often asked whether this means a species is newly minted by nature. It does not. Species are typically created slowly by evolution over many thousands or tens of thousands of years. Some species persist and are changed very little over hundreds of thousands, or even millions, of years. These species are sometimes called living fossils, such as the tiny Cyprus mouse that appears in chapter 3. In contrast, there is evidence of certain populations isolated on islands or mountaintops that seem to have fully completed a twelve-thousand-year process of evolving into a new species. That is blazingly fast for a new species to appear.

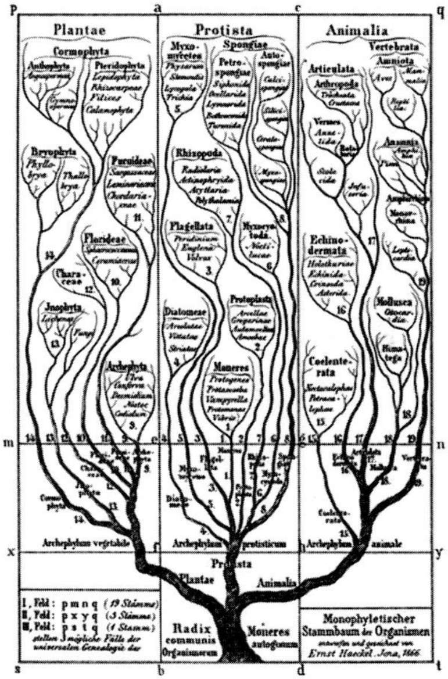

Many factors contribute to speciation—the evolutionary processes by which new species arise. Geographic isolation plays a major role. In many taxa, sexual selection is important, while other species arise by asexual means. Multiplication of chromosome numbers, a process called polyploidy, instantly isolates offspring from parent and is common among plants. And there are countless species, from insects to lizards, in which females reproduce in the absence of males through a process called thelytoky. In these organisms, sudden mutations instantly create new species and thereby account for many new species on our planet.

Ernst Haeckel's Tree of Life: *Generelle Morphologie der Organismen* (1866).

COURTESY OF WIKIMEDIA COMMONS

Advances in taxonomy have come as a result of greater access to all corners of the Earth. Isolated localities in dense jungles, on high mountaintops, and in deep sea trenches are becoming increasingly accessible to scientists, allowing species to be discovered in previously unexplored regions. Online access to geographic data provides clues about gaps in distributions and places where new species may exist, while satellite images allow for more precise targeting of promising habitats. Ironically, improved access is the result of the same activities that constitute threats to the survival of species, such as logging roads or slash-and-burn agriculture. In other cases, taxonomists are revisiting their assumptions about *where* species can exist. Some extreme environments where life was thought to be unlikely if not impossible have surprised us in recent years, turning up a wealth of new species. (More about this in chapter 8.)

There are also instances in which species are known to indigenous populations,

COURTESY OF VINTAGE PRINTABLE

COURTESY OF VINTAGE PRINTABLE

while remaining unknown in the scientific community. For example, a newly discovered yam has been a food source so greatly exploited by the locals in Madagascar that it is most likely now an endangered species. In these cases, the concept of "discovery" sometimes leads to charges of imperialism being leveled against the scientists from the industrialized world who get credit for discovering what has long been known to natives of the area. It is important to recall, however, that "discovery" in the scientific world indicates a species has been officially classified, including the critical practice of cladistic analysis discussed earlier. The orderly organization of species is more important than ever as we confront the rapid loss of species and degradation of Earth's ecosystems.

Scientists are often asked whether it is necessary or even desirable to complete an inventory of all the species on Earth. Human civilization has survived and more or less progressed for five thousand years in near complete ignorance of species, and so its "necessity" may appear tough to defend. While there are lots of reasons to explore species, three especially compelling ones are evolutionary curiosity, biodiversity sustainability, and biomimicry.

Evolutionary Curiosity

Charles Darwin explained how species come to be as a result of natural selection, but most interesting is the part that follows—understanding the actual history of the origins and transformations of species and the rise of those evolutionary novelties that make them unique. While we have often contemplated some species' conspicuous characteristics, like rhinoceros horns and giraffe necks, the biosphere is teeming with plants, animals, and microbes that are, in their own ways, no less improbably weird and wonderful. Each has an amazing story to tell us if we will just take the time to listen.

One thing that makes us human is our innate curiosity about ourselves, our origins, and our place in the universe. A critically important part of the answer lies in the complex story of evolution. As we piece together the history of species—as revealed by morphology, ecology, behavior, fossils, and the genome—we begin to appreciate our status as a species within evolutionary history. There is a continuum that links us to the first single-celled organism, and we can never truly be divorced from its long and amazing tale. Because we are now facing serious environmental shifts and changes, a huge number of species could soon disappear, leaving behind no record of themselves. Few gifts to posterity could be more important than building great museums so that future generations can continue to wonder, explore, and admire biodiversity.

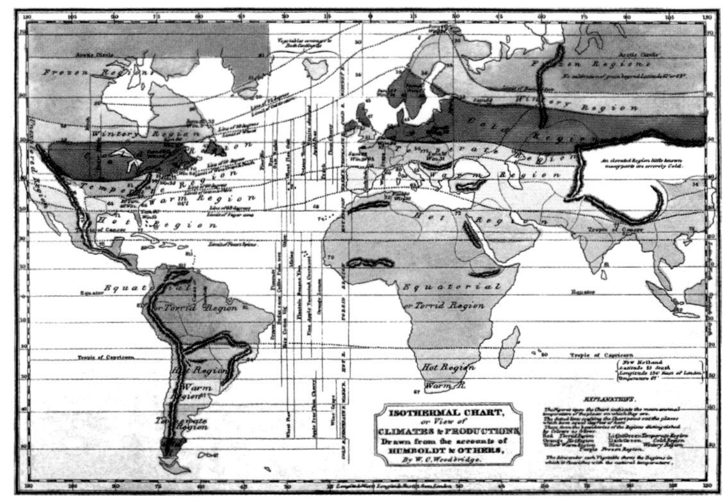

COURTESY OF VINTAGE PRINTABLE

Biodiversity Sustainability

A second reason to continue species exploration is biodiversity sustainability. All life on Earth—including the human species—depends upon healthy, resilient ecosystems to provide clean air, potable water, nutrient cycling, and carbon sinks (reservoirs that hold more carbon than they release). Clever engineers might build machines to deliver some of these services but this is a precarious alternative to nature.

Single species are not very resilient over the long haul. Emergence of a disease, a sudden change of climate, or some other stress can wipe them out. Few species even leave a fossil record behind although hundreds of new fossil species are found every year (chapter 5). Our best bet is sustainable biodiversity. Ecosystems that are diverse are more resilient to unexpected change and more likely to adapt to whatever the future throws at them. Unless we know what species already exist in an ecosystem, however, we are powerless to detect species loss or to monitor changes in biological diversity. Exploring species is the only way to face an uncertain environmental future with confidence.

Biomimicry

Mother Nature has had nearly 4 billion years to conduct endless, random, trial-and-error experiments. Most failed. Some have succeeded—and with spectacular results. As human society faces environmental challenges of unprecedented complexity, we can turn to biodiversity for solutions. We do not have the time to try countless alternatives, but luckily, evolution has already done the job for us. There are thousands of examples of designers, architects, material scientists, chemists, engineers, physicians, and others who have been inspired by nature with promising answers. Synthetic antibiotics are modeled after compounds that are produced by species for self-defense, such as penicillin. An office building in Zimbabwe uses passive air circulation modeled after a termite nest to reduce cooling costs by 90 percent. And new paints and fabrics mimic the microscopic surface structure of a plant's leaf to shed water in such a way as to be self-cleaning.

The possibilities are endless, yet we have only scratched the surface. How many clues for other solutions exist among the 10 million species we do not yet know?

Species: The Final Frontier

We have given reasons for continued species exploration to the curious, the desperately worried, and the profit seekers. That should pretty much cover everybody. We do not particularly care whether you are motivated by a need for intellectual stimulation or wild-eyed capitalist greed, so long as you join us in our mission to explore strange new worlds and seek out new life. As jazzed as we were by the unsuspected species discoveries at the beginning of this century, we anxiously await the hundreds of thousands of species that will be discovered in the years ahead.

Current advances in digital and computer technologies are in the process of changing everything about species exploration. In the next decade, taxonomists will have access to instrumentation, communications, data, and other research resources that will transform their science and vastly accelerate the rate at which species can be made known. The hundred species in the following chapters are just a teaser. The main act is just beginning.

The reader will no doubt wonder how we selected one hundred species from the nearly two hundred thousand named over the past decade. First, a confession—it was a whale of a task. There are, without doubt, thousands of candidate species that would have knocked off our collective socks had we been able to review over a million pages in scientific journals. With apologies, we can say only that we made countless substitutions and some tough choices and ultimately picked species that we personally found fascinating, surprising, sometimes disgusting, or just cool. In general, we were able to agree,

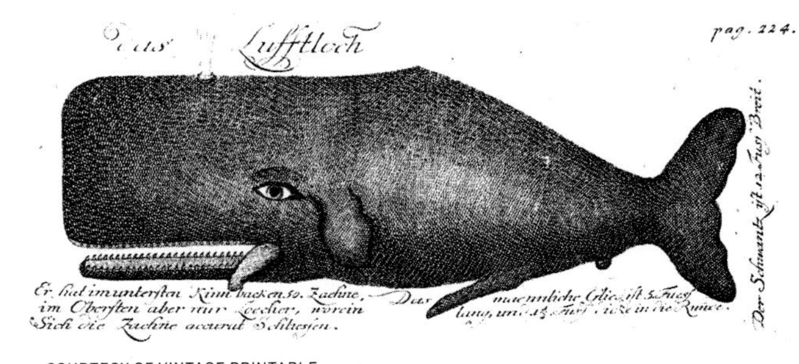

COURTESY OF VINTAGE PRINTABLE

but some made it into the book or were rejected only after a few loud conversations and unvarnished sharing of opinions. Each of us will tell you in private that we were right, but in the interest of world peace and editorial deadlines, we came to the compromise across the ten chapters that follow.

In the end, the choice has to be arbitrary on some level. After all, because every single species is unique in terms of its evolutionary history and its role in one or more ecosystems, they all deserve to be in this book. If you just look at any species closely enough, you will learn something truly remarkable about the biosphere, something you could never have dreamed to ask. Our role is that of a taxonomic tour guide, to share a few entertaining facts drawn from a mere one hundred stories out of the millions that exist. We hope that this will inspire you to encourage museums to expand and to support taxonomists in their discoveries. Better still, pick out your favorite taxon—that is, your favorite group of closely related species—and get your boots dirty, become an expert or enthusiast, and help us discover a few more of the 10 million species waiting to have their stories told.

WHAT ON EARTH?

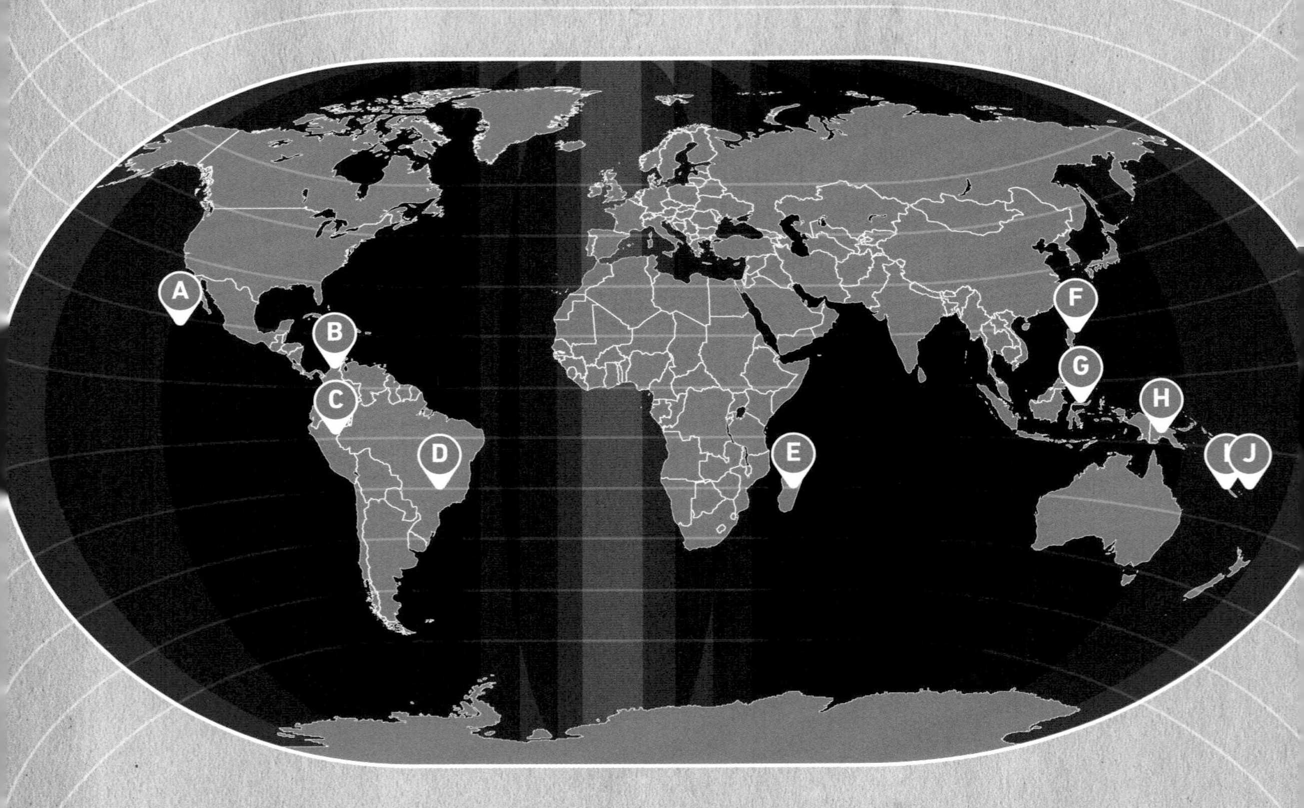

A Kaiser's Nudibranch

B Paisa Orchid Bee

C Kovach's Orchid

D Diamantina Taratula

E Patton's Bright Snake

F Fried Eggs Worm

G Psychedelic Frogfish

H Wilson's Blue-Eyed Cuscus

I Barbie Pagoda Fungi

J Exquisite Sea Urchin

Chapter 1

PRETTY COUNTS

THE PRETTIEST NEW SPECIES

Plant and animal species are so diverse that the old adage "beauty is in the eye of the beholder" could be the perfect slogan for nature's bounty. It's easy for most people to see the breathtaking beauty found in the brightly colored wings of butterflies, a field of blooming wildflowers, or a grove of hardwood trees in their autumn glory. But what about snails and their trails of slime, hoary rats with yellow teeth, or spiders that look like fierce aliens? These species are beautiful in their own right—just not in a traditional sense. Recognition of their unique beauty may require setting aside any preconceptions—or misconceptions—people may have about rotting fungi, creeping arthropods, or slithering snakes.

What we perceive as beauty is often due to form, coloration, or a combination of the two. People seem to be hardwired to see warm and fuzzy mammals as cute, while often lacking this innate and immediate attraction to the cold-blooded, eight-legged, egg-laying, or scale-clad members of the animal kingdom. Yet beauty is in no short supply among the invertebrates, the reptiles, or the fish, as several of our choices in this chapter will demonstrate.

It is worth noting that beauty can occur at all levels, from one to many and big to small, from individual organisms to entire ecosystems and back down again to microscopic cells and the beauty of the double helix of DNA. Layer upon layer of anatomical complexity presents opportunities to discover beauty in nature. The exquisiteness of a morpho butterfly's iridescent blue wing is even more amazing when viewed under a microscope. The surface of the wing consists of thousands of overlapping scales, each with minute grooves that diffract light and create its spectacular display of color. Similarly, a plant may be appreciated for its overall form or for bearing a beautiful flower. Looking more closely at the intricate structure of a flower's stamens and petals, another level of beauty can be observed. With even closer examination, we discover that individual grains of pollen may have ridges, bumps, and pits that are beautiful in a way that is completely unanticipated.

Beauty is also found in ensembles of individuals and mosaics of species. Consider the fluid magnificence of a flock of birds or school of fish in motion, the majesty of an intricately structured coral reef, or the awe inspired by a multilayered canopy of a tropical rain forest. Beauty is not limited to physical appearances such as shape and color, either. We only need to think of the songs of birds or the delicate webs spun by orb-weaving spiders to be reminded of this fact.

The beauty of the natural world has inspired many of the great works of art and literature. In everything from still-life paintings to Shakespearean sonnets, humans have tried to express our emotional responses to the beauty found in other species for thousands of years. Cave paintings, bone carvings, and other primitive art forms include depictions of deer, bison, and other species that not only documented animals familiar or important to early humans but may have also served to beautify our ancestors' dwellings. The influence of nature's beauty on artists is no less influential today, as a visit to any modern art museum will attest. As Yogi Berra once said, "You can observe a lot just by watching." This may be as simple as slowing down on a walk to admire the details of a mushroom cap's gills or looking up and stopping to snap a photo of a tree's branches.

Taking the time to *really* look at species is more often than not met with unsuspected and unexpected wonder. The spectacular diversity of nature can be as inspiring to the amateur admirer as it is to the scientists who devote their careers to the exploration of a single group of species, seeking to understand the evolution of wondrously complex, improbable, and beautiful anatomical structures. No matter where you are, and regardless of the organisms that surround you, the observer who is open to the wonder of nature will be rewarded with delights. By focusing on just a few of the pretty species newly discovered, we hope to renew, or awaken, your appreciation for beauty where you may not have expected to find it.

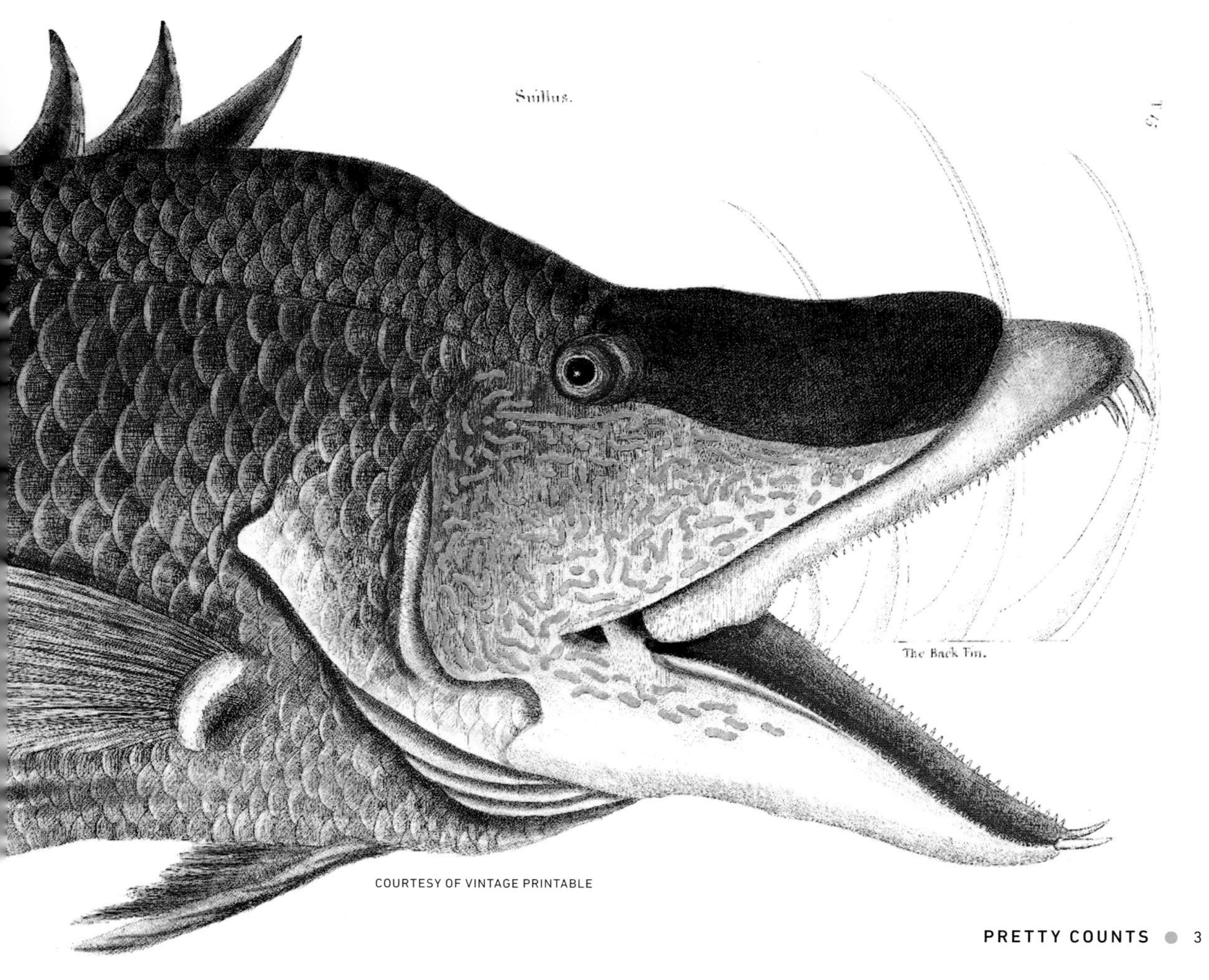

COURTESY OF VINTAGE PRINTABLE

ALICIA HERMOSILLO

Kaiser's Nudibranch

NUDE TO SCIENCE

Often found swimming in underwater caves off the west coast of Mexico, the new species *Polycera kaiserae* is a wonderful example of how beautiful and diverse nudibranchs can be. Although nudibranchs in general are known for their striking jewel-like colors, *P. kaiserae* is particularly distinctive for its white polka dots on a pink body with navy blue tips. The word "nudibranch" is from the Latin word *nudus* meaning nude and the Greek word *brankhia* meaning gills. Scientists believe that over time, nudibranchs, often called sea slugs, evolved to shed their outer shells and develop two defense mechanisms: camouflage to hide from predators and the ability to release acid from their skin to become distasteful. Their amazingly bright and beautiful colors not only make them noticeable but also protect them through "aposematism," an evolutionary adaptation in which distinct and conspicuous coloration becomes a warning to potential predators. There are over three thousand known species of nudibranchs around the world.

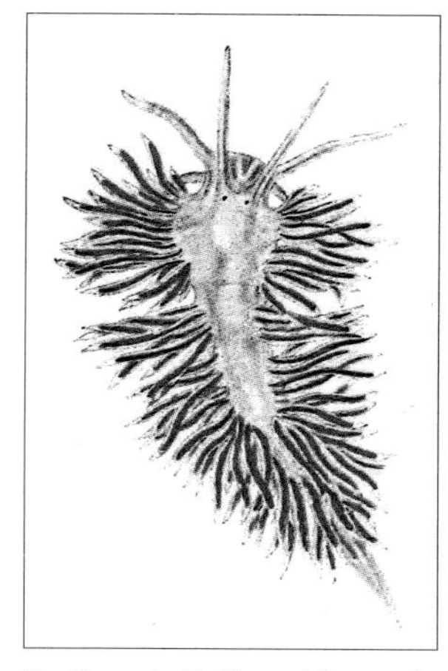

Nudibranch, *Nudibranchfaunaeni Drøbaksundet* (1922).

COURTESY OF VINTAGE PRINTABLE

DISCOVERED	Pacific coast of Mexico
SCIENTIFIC NAME	*Polycera kaiserae* Hermosillo & Valdés 2007
ETYMOLOGY	*Polycera* is a genus of nudibranchs named in 1817; *kaiserae* is named in recognition of Kirstie Kaiser's dedication and contributions to the molluscan fauna of the eastern Pacific Ocean.
CLASSIFICATION	Animalia • Mollusca • Gastropoda • Nudibranchia • Polyceridae

SANTIAGO RAMIREZ

Paisa Orchid Bee

SAY *AAHHHHHH*

SANTIAGO RAMIREZ

As pretty as the orchids they pollinate, orchid bees come in a brightly colored array of iridescent blues, greens, and purples. These bees are not social like honeybees—they are typically solitary in nesting, with no division of labor and little communal activity. The males leave the nest shortly after birth and never return, spending their lives collecting flower fragrances that they store in special grooves on their hind legs and that may be released to attract females. Females construct nests from mud, resins, and other materials and gather both nectar and pollen from a variety of plants. The two hundred species of orchid bees are native to Central and South America and play an important role in the pollination of many orchids. Males of *Euglossa paisa* (Paisa orchid bee) have tongues that are four times longer than their heads and approximately the same length as their entire body, a valuable asset to pollination. The relationship between orchid bees and orchids is a varied and complex one. In the most extreme cases, orchids have adaptations that precisely place their pollen packets on male bees in a way that ensures they will be delivered only to another flower of the same species.

DISCOVERED La Forzosa Natural Reserve in Antioquia, Colombia

SCIENTIFIC NAME *Euglossa paisa* Ramírez 2005

ETYMOLOGY *Euglossa* is a genus of orchid bees named in 1802; *paisa* is in recognition of the Paisa region, where the species is found.

CLASSIFICATION Animalia • Arthropoda • Insecta • Hymenoptera • Apidae

Kovach's Orchid

LAW AND ORCHID

Smuggling and scheming, international intrigue and scandal—the discovery of Kovach's orchid had more ploys than a whodunit. Perhaps one of the most spectacular orchids of the century, with its amazing color and size (almost eight inches from tip to tip), *Phragmipedium kovachii* was purchased by an orchid entrepreneur in 2002 at a roadside flower stall in rural Peru. Excited that the plant was most likely a new species, the orchid hunter smuggled his purchase out of Peru to a botanical garden in the United States. Scientists determined that the flower was indeed a new species, and its description was rushed to press. But, like the bank robber who leaves his name on a deposit slip, the collector (for whom the species is named) and the botanical garden (which published the description in its own journal) were soon caught orchid-handed and charged with the federal crime of illegally importing and possessing an endangered species. *P. kovachii* remains in the protective custody of the United States Botanic Garden.

© DR. HENRY OAKELEY

DISCOVERED	Northern Peru
SCIENTIFIC NAME	*Phragmipedium kovachii* Atwood, Dalström & Fernandez 2002
ETYMOLOGY	*Phragmipedium* is a genus of lady's slipper orchids named in 1896; *kovachii* after J. Michael Kovach, who illegally transported the orchid from Peru.
CLASSIFICATION	Plantae • Angiosperms • Monocots • Asparagales • Orchidaceae

ADNAN FARHAT

Diamantina Tarantula

ROSES ARE RED, TARANTULAS ARE BLUE?

Even the world's biggest arachnophobes would have to admit that *Oligoxystre diamantinensis* (Diamantina tarantula) is one of this century's most beautiful new species to be described by scientists. *O. diamantinensis* is neither the first nor the most recently discovered blue tarantula but it's certainly one of the most spectacular. Interestingly, its blue metallic color is not lost in preservation ethanol, indicating that the spider's coloration and pigmentation are not due to diet or environmental factors. Like other tarantulas, this blue beauty does not have venom that is lethal to humans, although allergic reactions are potentially dangerous.

ROGERIO BERTANI

DISCOVERED	Minas Gerais, Brazil
SCIENTIFIC NAME	*Oligoxystre diamantinensis* Bertani, dos Santos & Righi 2009
ETYMOLOGY	*Oligoxystre* is a genus of tarantulas named in 1924; *diamantinensis* is for the city and montane savanna ecoregion of Diamantina, where the species was discovered.
CLASSIFICATION	Animalia • Arthropoda • Arachnida • Araneae • Theraphosidae

DAVID R. VIEITES

Patton's Bright Snake

MADAGASCAR COLOR MADNESS

Located in the Indian Ocean off the southeastern coast of Africa, the island country of Madagascar is a biodiversity hot spot. The fourth largest island in the world, Madagascar separated from India about 90 million years ago and as a result of its long isolation, has species found nowhere else on Earth. Three or four families of snakes live on Madagascar, and the crayon-colored pattern of *Liophidium pattoni* is unique among the island's ninety known snake species. Bright coloration is often a warning to predators that a species is poisonous or foul tasting. However, *L. pattoni* appears to be neither aggressive nor lethal. Its dorsal (back) pattern of red spots on black may be mimicking a poisonous species of millipede that has a similar color pattern. Although ophidiophobia (fear of snakes) is fairly common, it may be hard to resist this gorgeous guy.

DAVID R. VIEITES

COURTESY OF VINTAGE PRINTABLE

Crocodile or caiman and South American false coral snake, Maria Sibylla Merian (c. 1705–1710).

DISCOVERED	Makira Natural Park, province of Mahajanga in northeastern Madagascar
SCIENTIFIC NAME	*Liophidium pattoni* Vences, Vieites, Ratsoavina & Randrianiaina 2010
ETYMOLOGY	*Liophidium* is a genus of snakes named in 1896; *pattoni* after renowned mammalogist Jim Patton.
CLASSIFICATION	Animalia • Chordata • Reptilia • Squamata • Colubridae

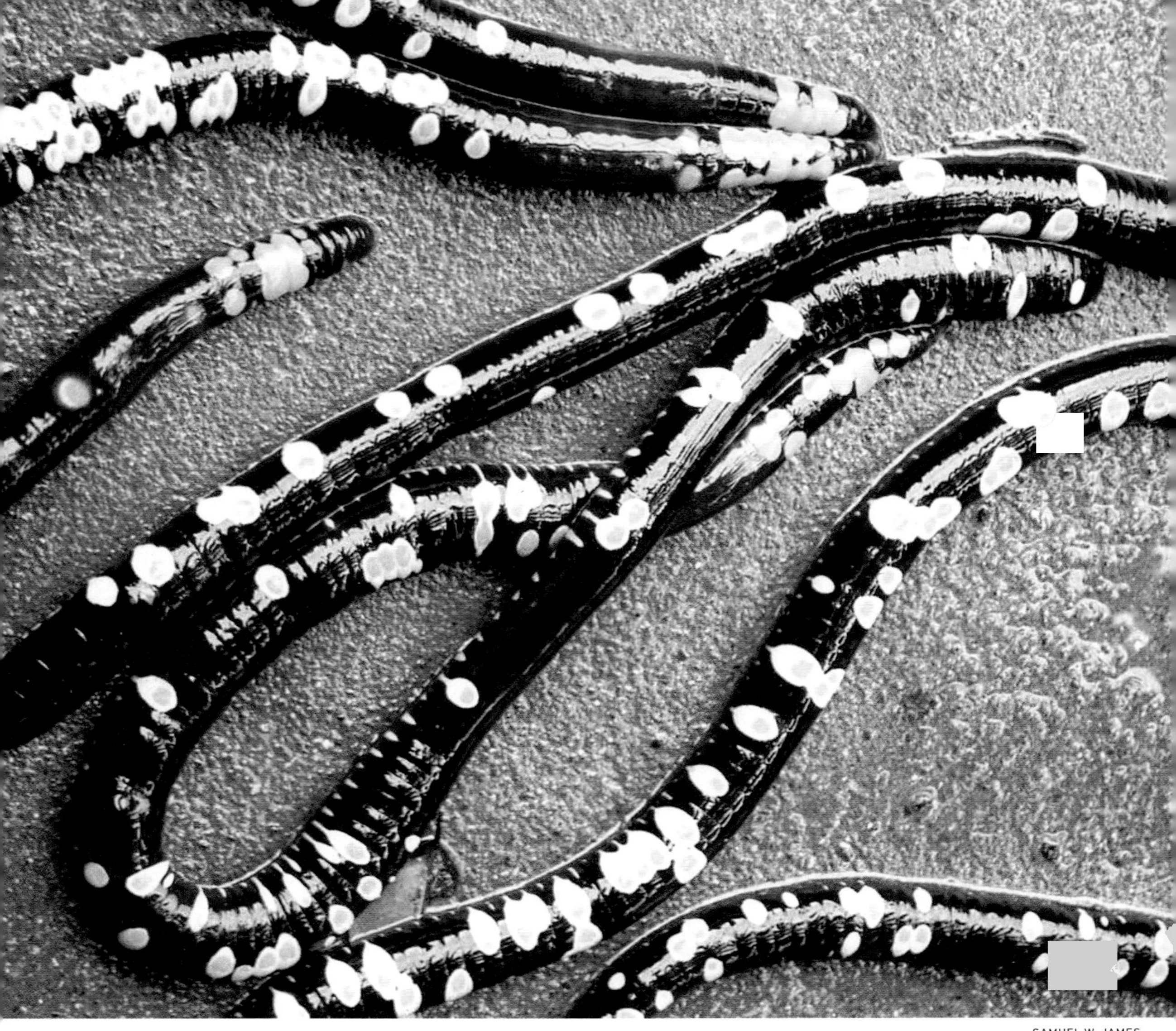

SAMUEL W. JAMES

Fried Eggs Worm

OVER EASY, EASYGOING

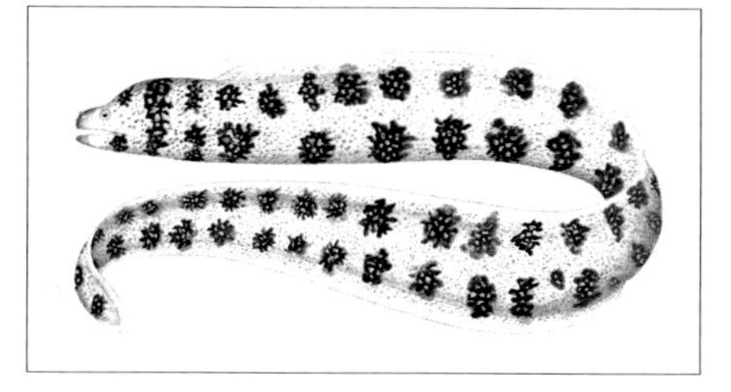

Snowflake moray eel (*Echida nubulosa*) by A. H. Baldwin in *The Bulletin of the United States Fish Commission* (1903).

First found in 2001 in the old-growth forest of the Philippines' Maria Aurora National Park, this new species of worm is remarkable not only for its beautiful coloration but also because it doesn't burrow into the soil like most earthworms. Instead, it calmly moves about in the open on the forest floor with a laid-back attitude that would make a California surfer proud. According to the scientists who discovered *Archipheretima middletoni*, its behavior was so calm during handling that it didn't even squirm to escape and would remain still during its photography sessions. What a ham! Although all species of the genus are blue, this pretty worm is the only one to have the yellow and white spots that resemble fried eggs. These spots most likely create protective camouflage effects either by looking like patches of reflecting light on leaf litter or by making the outline of the worm's body difficult for predators to see. Local folklore includes the belief that these large, easygoing worms travel to the nearby river, where they turn into the river's large spotted eels.

DISCOVERED Philippines

SCIENTIFIC NAME *Archipheretima middletoni* James 2009

ETYMOLOGY *Archipheretima* is a genus of earthworms named in 1928; *middletoni* is named in honor of Robert Hunter Middleton, an American designer and calligrapher from Chicago, Illinois.

CLASSIFICATION Animalia • Annelida • Clitellata • Haplotaxida • Megascolecidae

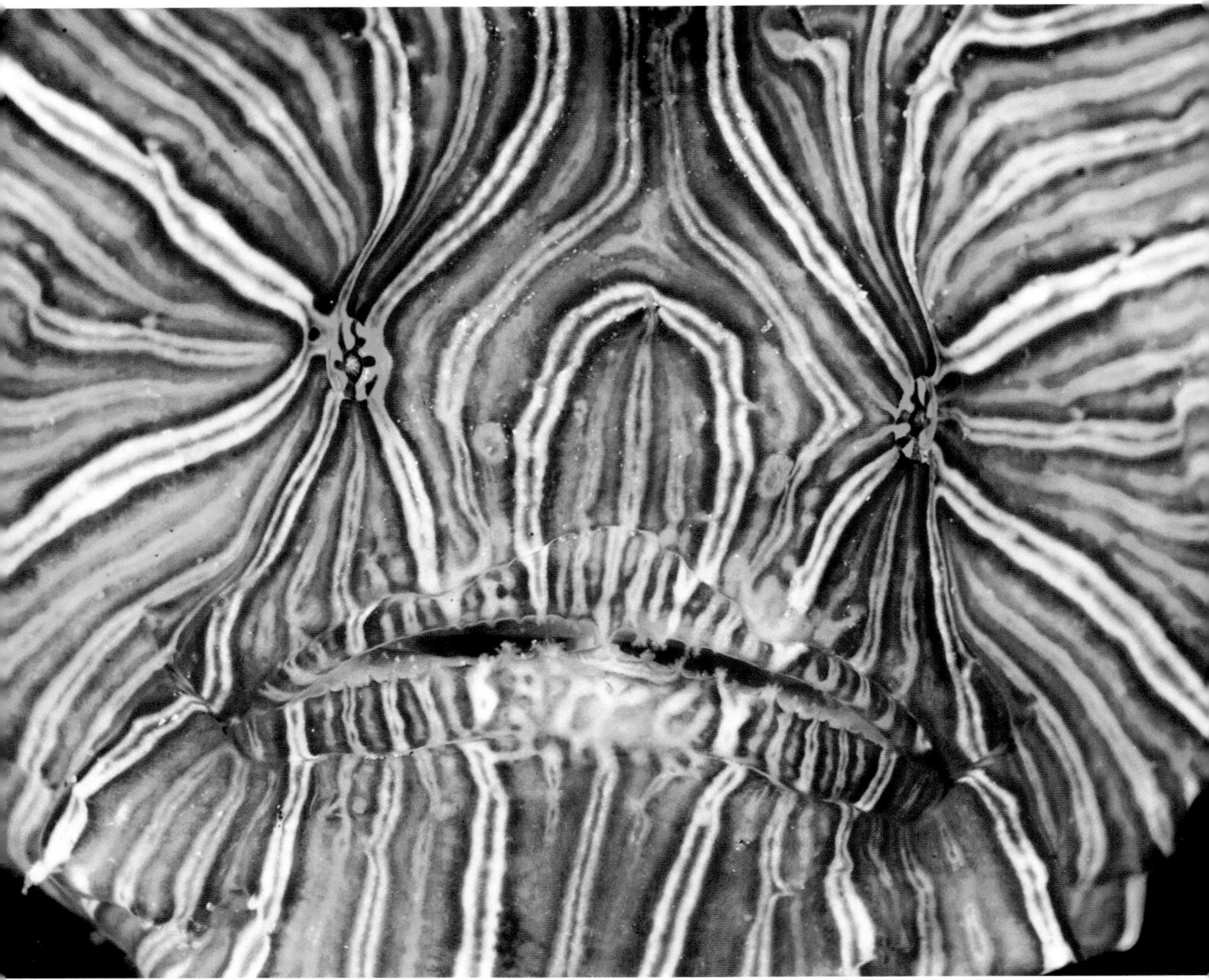

DAVID HALL / SEAPHOTOS.COM

Psychedelic Frogfish

LIFE IMITATING ART?

With public installations around the world in hotels, museums, and botanical gardens, Puget Sound artist Dale Chihuly is renowned for creating magnificent glassworks that stretch the imagination with their form, color, and breathtaking beauty. When *Histiophryne psychedelica* was discovered in 2008, the University of Washington press officers were so smitten that they described this new frogfish as having "a face even Dale Chihuly could love." In a case where life may be imitating art, *H. psychedelica* is either one of the most bizarre or one of the most beautiful fish to have been described this century. With its amazing coloration of radiating, swirling stripes and a wide, flat face that emphasizes its turquoise eye shadow, this is definitely one of nature's prettiest new species. Its unique pattern of pigmentation is believed to provide camouflage protection in the coral reefs where it lives.

© DAVID HALL / SEAPHOTOS.COM

Journal des Museum Goddeffroy (1873).

DISCOVERED	Ambon and Bali, Indonesia
SCIENTIFIC NAME	*Histiophryne psychedelica* Pietsch, Arnold & Hall 2009
ETYMOLOGY	*Histiophryne* is a genus of frogfish named in 1863; *psychedelica* to reflect its abstract and mind-blowing multicolored patterns of stripes and swirls.
CLASSIFICATION	Animalia • Chordata • Actinopterygii • Lophiiformes • Antennariidae

TIM FLANNERY

Wilson's Blue-Eyed Cuscus

I'VE GOT MY EYE ON YOU

TIM FLANNERY

Possums, not to be confused with the New World opossums, are a group of about seventy marsupials found only in Australia, New Guinea, Sulawesi, and nearby islands. These small- to medium-sized animals live in trees and primarily hunt for food at night. One of the most recently described cuscuses, *Spilocuscus wilsoni*, is also among the most threatened. *S. wilsoni* was not recognized as a new species until 2004, although a specimen had been collected in 1906 and deposited in the Netherlands Center for Biodiversity. In 1992, scientists photographed *S. wilsoni* living as a family pet on the island of Biak in Papua, New Guinea, and noted it was most likely an animal new to science. *S. wilsoni* may be the smallest of the spotted cuscuses, and it is the only species in its genus that has extraordinary blue-green eyes, making it one of the prettiest animals to be described this century. This species is extremely rare and may be critically endangered—it was not seen in the wild during an extensive mammal survey conducted on the island in 1992 nor was it collected during major scientific expeditions in 1962 and 1976.

DISCOVERED Papua, Indonesia

SCIENTIFIC NAME *Spilocuscus wilsoni* Helgen & Flannery 2004

ETYMOLOGY *Spilocuscus* is a genus of marsupials named in 1862; *wilsoni* is named in honor of Don E. Wilson of the National Museum of Natural History in Washington, D.C.

CLASSIFICATION Animalia • Chordata • Mammalia • Diprotodontia • Phalangeridae

MARC DUCOUSSO

Barbie Pagoda Fungus

PRETTY IN PINK

Barbie may have more pink cars than she can drive and a pink three-story town house, but the Pacific island archipelago of New Caledonia has *Podoserpula miranda*, a beautiful candy pink fungus structured like a multi-tiered tower. Nicknamed Barbie Pagoda, *P. miranda* not only has a highly unusual color for a fungus but also an architecture that is especially fascinating to scientists, containing a unique set of six mushroom caps that are centered on one stalk and decrease in size as they near the top. Approximately three hundred species of New Caledonia fungi are known to science, and researchers estimate there are as many as thirty thousand fungus species on the island that are yet to be discovered and described. Sadly for Barbie, *P. miranda* is only one-third of her height, so it's doubtful she'll be adding this architectural wonder to her array of pink residences.

MARC DUCOUSSO

DISCOVERED	New Caledonia
SCIENTIFIC NAME	*Podoserpula miranda* Ducousso, Proust, Vigier & Eyssartier 2009
ETYMOLOGY	*Podoserpula* is a genus of fungi named in 1963; *miranda* from the Latin, meaning wonderful.
CLASSIFICATION	Fungi • Basidiomycota • Agaricomycetes • Agaricales • Amylocorticiaceae

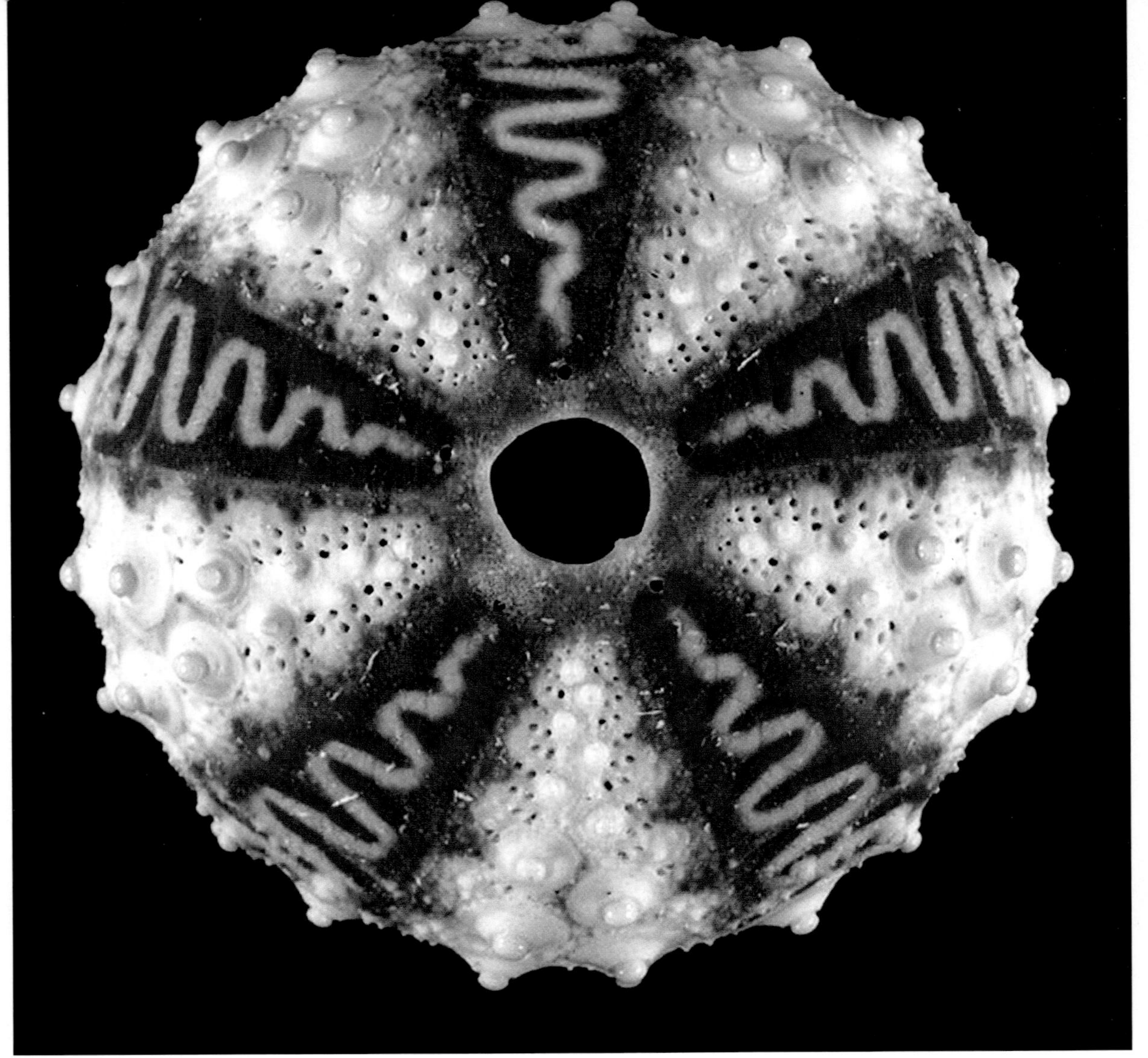

SIMON E. COPPARD

Exquisite Sea Urchin

FIVEFOLD NAKED BEAUTY

SIMON E. COPPARD

Almost a third of a mile beneath the surface of the South Pacific lives the exquisitely colored new species *Coelopleurus exquisitus*. Most sea urchins are commonly found in dull shades, but *C. exquisitus* is distinctive for having curved red and pale green banded spines and five naked areas that are raspberry pink with squiggly lines of lavender (the interambulacral median regions). As they develop, regularly shaped sea urchins change from a body with two mirror-image halves to a symmetrical body that is divided into five parts that equally radiate from the center. This fivefold radial symmetry, called pentamerism, can also be found in sea stars (starfish) and many flowering plants—even horizontally cut apples. *C. exquisitus* is also distinctive for its big mouth (the peristome), which is 56 percent of its body's horizontal diameter excluding the spines.

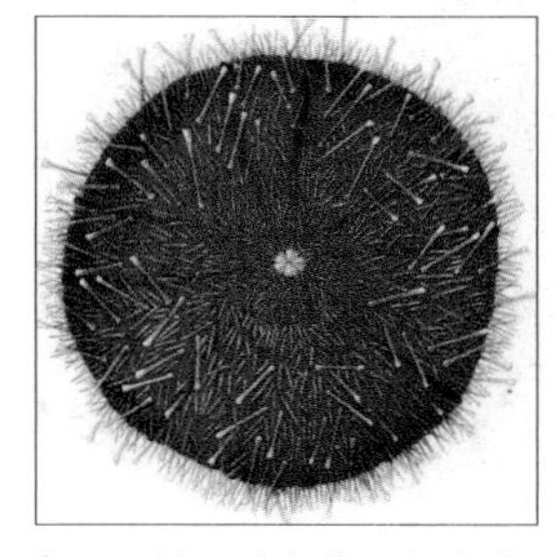

Sea urchin, original zoological drawings and plates, Museum of Comparative Zoology, Harvard University (1830–1880).

COURTESY OF VINTAGE PRINTABLE

DISCOVERED	New Caledonia
SCIENTIFIC NAME	*Coelopleurus exquisitus* Coppard & Schultz 2006
ETYMOLOGY	*Coelopleurus* is a genus of sea urchins named in 1840; *exquisitus* is named after the species' exquisite coloring that includes raspberry pink, red, green, orange, and white. Sea urchins are members of the Echinodermata phylum, which is based on the Greek word *echinodermate* meaning "spiny skin." The word "urchin" is from the Middle English word *urcheon*, which was used as a term for hedgehog, a spiny mammal that sea urchins are thought to resemble.
CLASSIFICATION	Animalia • Echinodermata • Echinoidea • Arbacioida • Arbaciidae

- A Little Fork Orchid
- B Quechuan Broad-Nosed Bat
- C Gomes's Freshwater Stingray
- D Double-Hooked Anglerfish
- E Dumbo Octopus
- F Big Brain Protist
- G Little Grooves Earthstar
- H Long-Neck Assassin Spider
- I Sahyadri Nose Frog
- J Ausubel's Mighty Claw Lobster

Chapter 2

STRANGER THAN (SCIENCE) FICTION

THE STRANGEST NEW SPECIES

When it comes to strange, even the most imaginative science fiction writers and Hollywood special effects artists have nothing on Mother Nature. This chapter includes just a sampling of the ugly, bizarre, and just plain weird species that have been discovered in the twenty-first century. Strangeness, like beauty, is a matter of opinion and also depends on context. Some of the species we've selected, such as the earthstar and leaf insect, are fairly typical of the groups to which they belong, yet, in general, share properties or appearances that are wildly unusual in the context of plants and animals. Others are utterly unique. All deserve attention for the wonderful contributions they make to biodiversity.

Most of our choices were made visually—that is, the plants and animals in this chapter simply look strange or weird. In many cases, strange looks are accompanied by equally strange behaviors or environmental circumstances. For example, the odd, elongated appendage over the head of the anglerfish looks bizarre, but more than that, the fish uses it to attract prey by bobbing and dangling it like bait on a fishing pole.

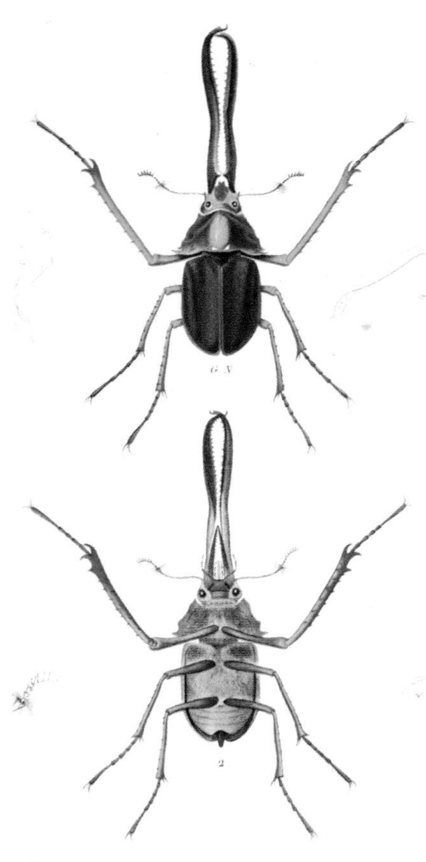

COURTESY OF VINTAGE PRINTABLE

Natural selection has picked winners and losers in the competition for survival across the 3.8 billion years that life has existed on our planet. In the process, plants and animals have evolved countless adaptations to challenges large and small. Even when an adaptation seems self-explanatory, such as an insect mimicking a leaf in order to camouflage itself, the details favored by natural selection are diverse and often surprising. Many insects seem to do just fine by simply being green and blending in colorwise. We do not yet know whether these imitations are good enough to fool all of an insect's particular predators, but we are quite certain that an insect evolving detailed textures and contours of a leaf is strange.

Scientists study the structure and form of plants and animals for many reasons. Comparing the body parts and shapes of species allows us to accomplish a lot: to recognize species by eye in the field; to explore the ways in which organisms have adapted to their environments; to reconstruct the evolutionary stories of how ancestral structures have been modified in descendant species; and to have a lot of fun in the process. Finding a new variation on a body part or a wholly new structure is fantastically exciting, like turning the page in a fast-paced mystery. Because anatomy and behavior lie at the interface between an organism and its environment, a species' appearance is often molded by natural selection. One result is a seemingly astronomical number of variations on morphology.

There is one additional reason that we should explore biodiversity and try to understand the many ways in which species have evolved strange and peculiar shapes, forms, and functions. Like every species before us, we live in a dynamic world. The surface of the Earth, and especially its biosphere, is undergoing constant change. In human timescales, that change is sometimes slow and gradual, as evidenced by the cycles between ice ages or the drifting of continents. Other times, change is sudden and violent, such as a volcanic eruption, forest fire, or tsunami. The geologic and real-time lesson is that we, like all life, must be prepared to adapt to our changing planet if we are to survive and prosper.

By studying and documenting new species, we can literally find billions of clues and answers to the challenges we face. Better materials, designs, and ideas are all around us in the structure and adaptations of plants, animals, and microbes—especially among the strange ones. We have only begun to crack this adaptation code and more than 90 percent of the best hints are yet to be discovered.

COURTESY OF VINTAGE PRINTABLE

DIEGO BOGARÍN

Little Fork Orchid

BUD-UGLY FLOWER

Ah, the language of flowers. In Victorian times, the kind of flower you gave someone had its own message—a yellow tulip meant "hopeless love" and a rose without thorns meant "love at first sight." Orchids were typically expressions of "beauty" and "refinement," so this new species would have sent a mixed message—even its discoverers described it as "bizarre," which is hardly a synonym for pretty. The amazing floral structures of Earth's twenty-five thousand known orchids are simply adaptations to attract and use insect pollinators. Across all colors, sizes, and sometimes questionable aesthetics, orchids will continue to yield amazing finds for generations to come. The Little Fork Orchid is less than half an inch wide and was discovered growing at an altitude of four thousand feet in the cloud forests of Panama's mountains near Cerro Arizona.

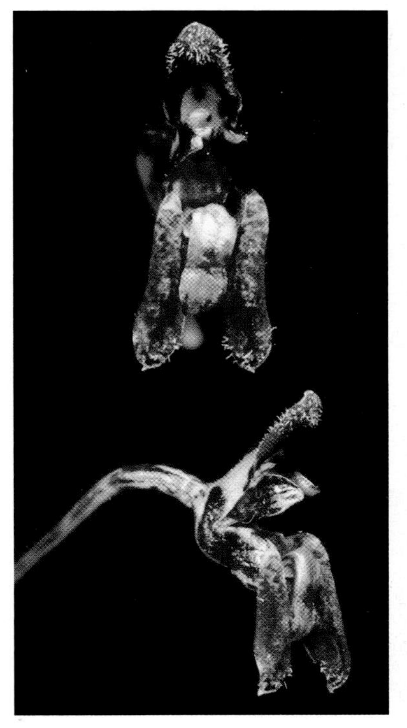

DIEGO BOGARÍN

DISCOVERED Panama

SCIENTIFIC NAME *Condylago furculifera* Dressler & Bogarín 2007

ETYMOLOGY *Condylago* is a genus of orchids named in 1982; *furculifera* is from the Latin word *furcula*, meaning "little fork," to describe the forklike stems that are attached to the flowers.

CLASSIFICATION Plantae • Angiosperms • Monocots • Asparagales • Orchidaceae

EDITH ARIAS ARONE

Quechuan Broad-Nosed Bat

A NEW LEAF

EDITH ARIAS ARONE

BRUCE D. PATTERSON

Whenever Hollywood wants to make a nighttime scene really scary, they bring out the bats—flying and flitting, dodging and biting, and always getting into someone's hair. Yet the misunderstood bat is actually an ecological wonder. Insect-eating bat species (insectivores) drastically reduce the need for harmful chemical insecticides to control insect populations, and the fruit-eating species (frugivores) help pollinate flowers and spread plant seeds. Bats represent about one-fifth of the world's mammal species and they are among the most frequently discovered new mammal species each year. About twenty years ago, several broad-nosed bat specimens that had been collected during expeditions to the Andes in South America and stored in museums as *Platyrrhinus helleri* were later found to have been misclassified. After a close reexamination of the specimens this past decade, the number of known *Platyrrhinus* species doubled from ten to twenty, including the discovery of *P. masu* in 2005. Like many *Platyrrhinus* bats, this new species has a leaflike nose and a face that only a mother could love.

DISCOVERED the Cultural Zone of the Manu Biosphere Reserve in the province of Paucartambo, Peru

SCIENTIFIC NAME *Platyrrhinus masu* Velazco 2005

ETYMOLOGY *Platyrrhinus* is a genus of broad-nosed or leaf-nosed bats named in 1860; *masu* is from the Quechua word for bat—Quechua is the name of the language and people who live in the central region of the Andes Mountains.

CLASSIFICATION Animalia • Chordata • Mammalia • Chiroptera • Phyllostomidae

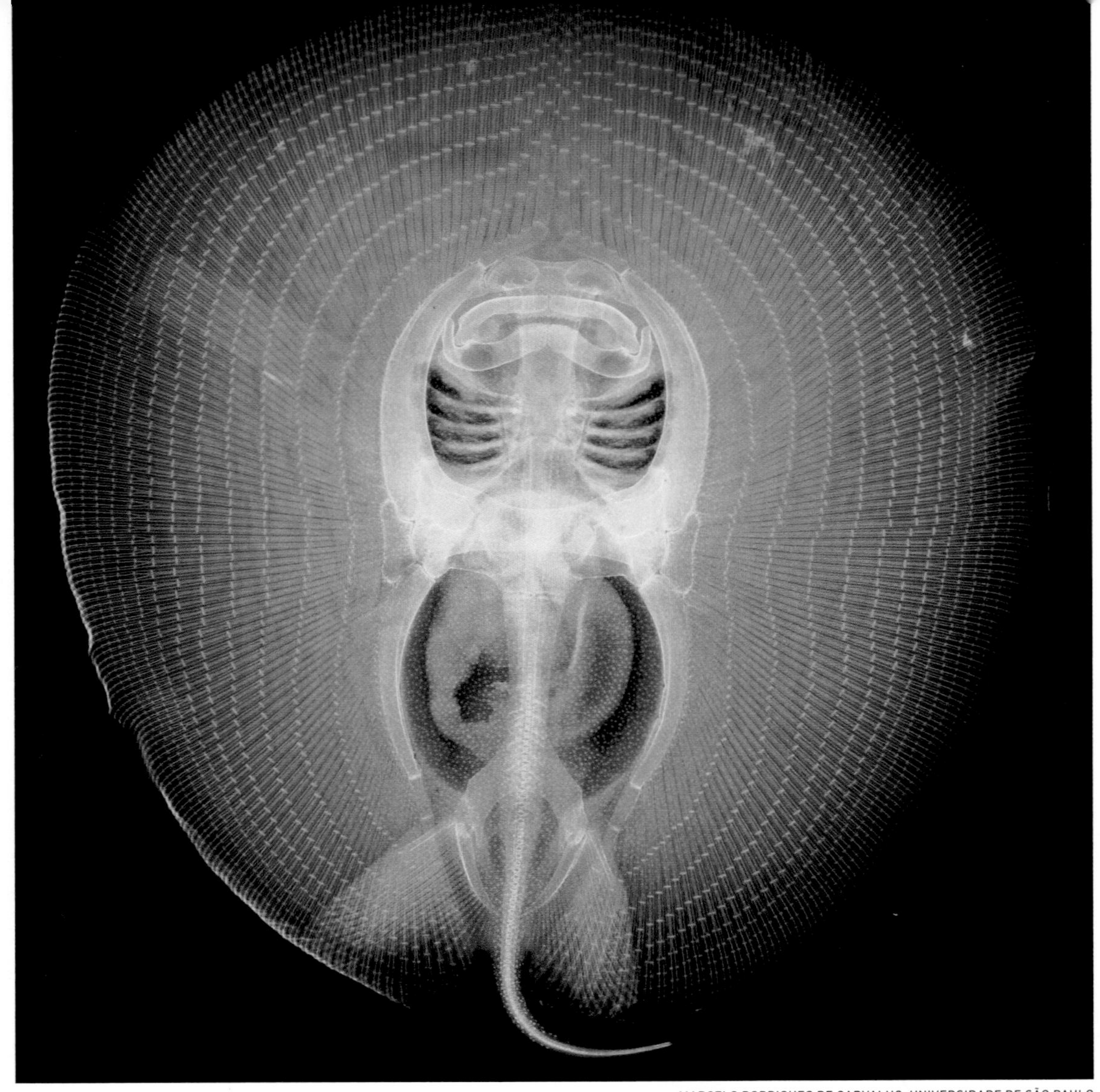

MARCELO RODRIGUES DE CARVALHO, UNIVERSIDADE DE SÃO PAULO

Gomes's Freshwater Stingray

X-RAY'D RAY

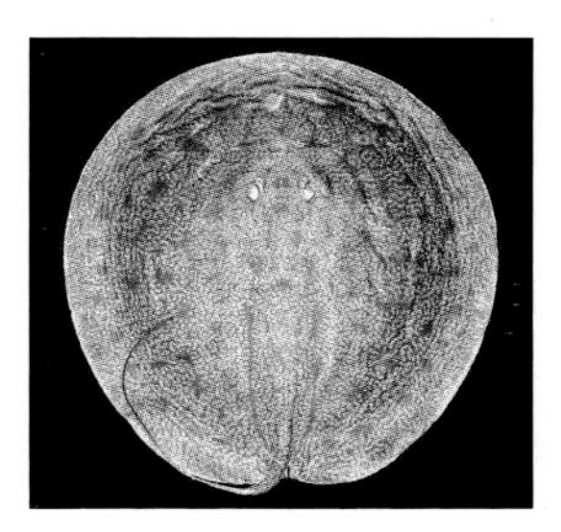

Heliotrygon gomesi, new species

MARCELO RODRIGUES DE CARVALHO, UNIVERSIDADE DE SÃO PAULO

When people think about stingrays, they generally imagine something big and flat flapping through tropical seas with their large triangular fins and long, stinging tails that can sometimes kill unsuspecting humans. This is not the case, however, for the freshwater stingrays in the family Potamotrygonidae that live in South American rivers and have amazingly round bodies (disks), short tails, and coloration that would make an animal print aficionado swoon. For more than a century, scientists have conducted field studies on the freshwater stingrays living in the Amazon River system and have discovered more than a dozen species, including the new genus and species *Heliotrygon gomesi*. Sometimes called a "round" or "china" ray because it looks like a big, round dinner plate, *H. gomesi* is especially unique for its reduced or absent caudal sting (tail); tiny, unprotruding eyes; and extremely circular body that can best be seen when its skeletal structure is x-rayed.

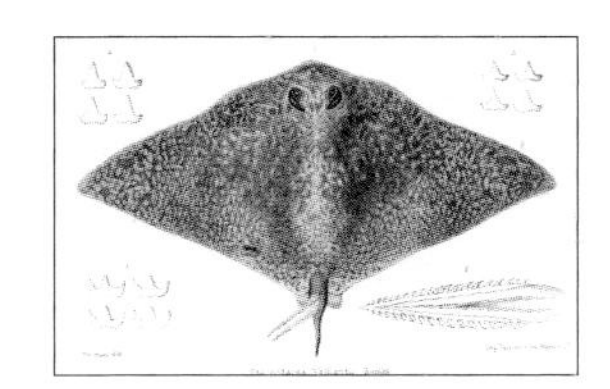

Stingray *Pteroplatea*, *Faune de la Sénégambie* (1883).

COURTESY OF VINTAGE PRINTABLE

DISCOVERED	Rio Jamari, Brazil
SCIENTIFIC NAME	*Heliotrygon gomesi* De Carvalho & Lovejoy 2011
ETYMOLOGY	*Heliotrygon* is a new genus of river stingrays named from the Greek words *helios* for sun, in reference to the ray's very circular body that appears to radiate outward, and *trygon*, meaning stingray; *gomesi* is in honor of Ulisses L. Gomes for his pioneering work with elasmobranchs.
CLASSIFICATION	Animalia • Chordata • Chrondrichthyes • Myliobatiformes • Potamotrygonidae

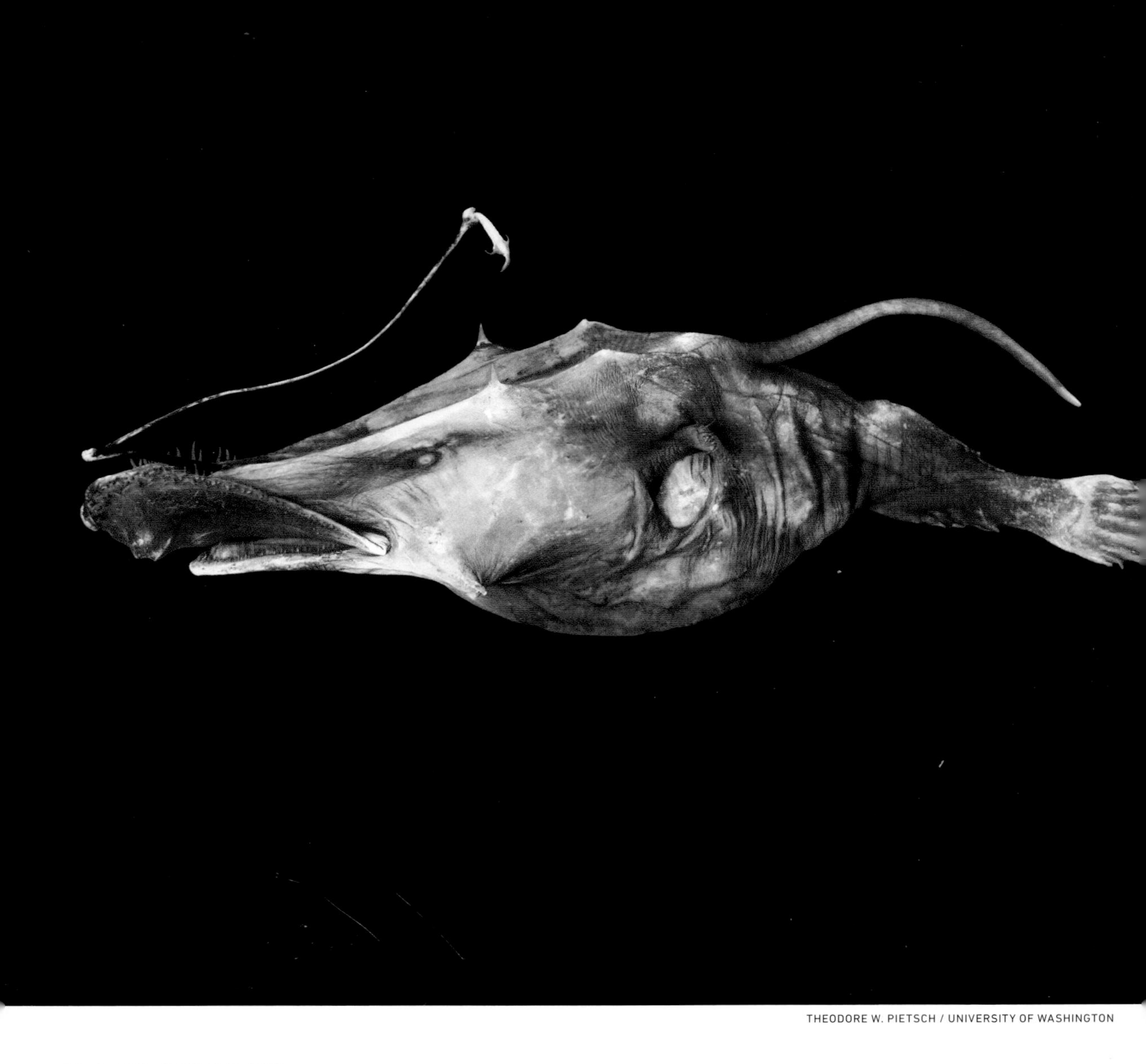

THEODORE W. PIETSCH / UNIVERSITY OF WASHINGTON

Double-Hooked Anglerfish

FLASHLIGHT TAG

Only the fifth species of deep-sea anglerfish of the genus *Lasiognathus* to be discovered, *L. amphirhamphus* is a good poster child for weirdness. With a head that takes up more than half its length; an enormous mouth; long, slender hooked teeth; and, of course, the esca—an elongated growth on the head that is dangled and wiggled like live bait on a line to attract prey—the double-hooked anglerfish is just plain bizarre. You couldn't make up a story as unique as this fish's lure. Like a fishing pole, a structure called the illicium extends from the first dorsal fin spine and at its tip is the fleshy elongated "bait." When the bait makes contact with unsuspecting prey, the fish's jaws are automatically triggered like a deadly trap and the prey is captured. The tip is also home to special bacteria that emit light to help attract prey in the total darkness of deep-sea waters—like waving a flashlight to get their attention. *L. amphirhamphus* was discovered at a depth of some four thousand feet in the Madeira Abyssal Plain off the coast of Madeira Island.

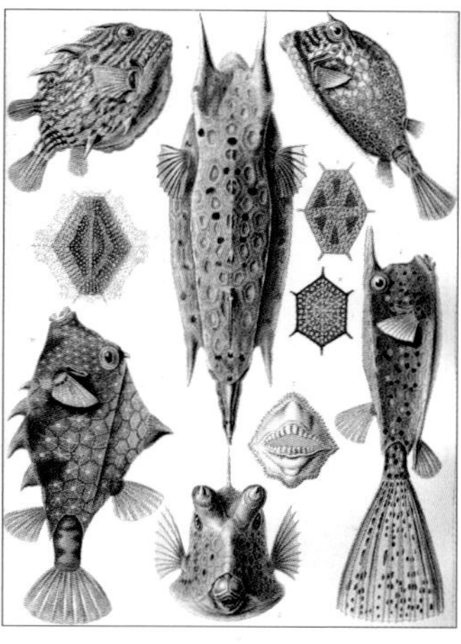

Boxfish, from Ernst Haeckel's *Art Forms of Nature, Kunstformen der Natur* (1904).

DISCOVERED	eastern central Atlantic Ocean
SCIENTIFIC NAME	*Lasiognathus amphirhamphus* Pietsch 2005
ETYMOLOGY	*Lasiognathus* is a genus of fish named in 1925; *amphirhamphus* is from the Greek words *amphi* meaning double or "both sides," and *rhamphus* meaning hook.
CLASSIFICATION	Animalia • Chordata • Actinopterygii • Lophiiformes • Thaumatichthyidae

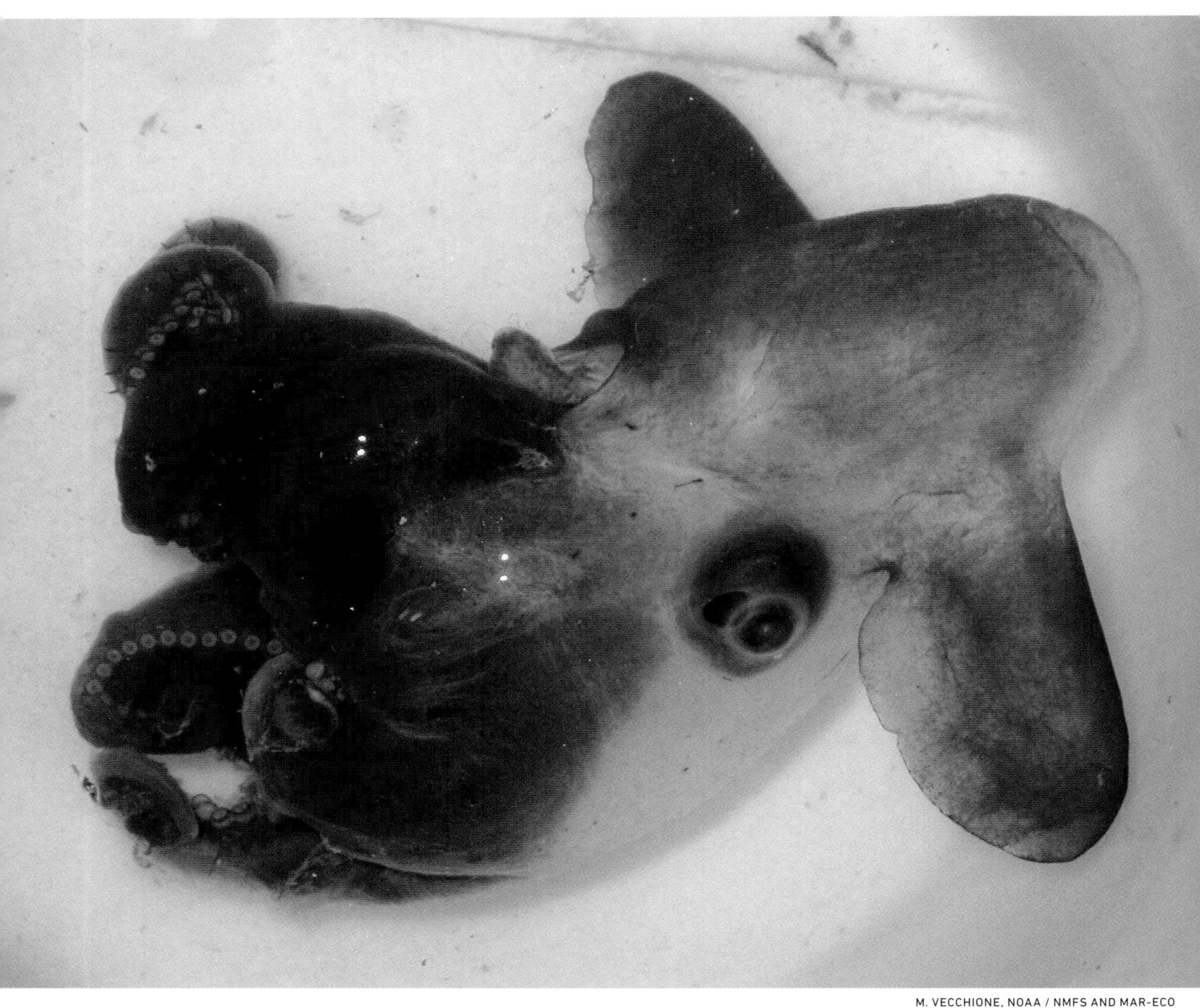

M. VECCHIONE, NOAA / NMFS AND MAR-ECO
PROJECT OF CENSUS OF MARINE LIFE

Dumbo Octopus

EARS LOOKING AT YOU, KID

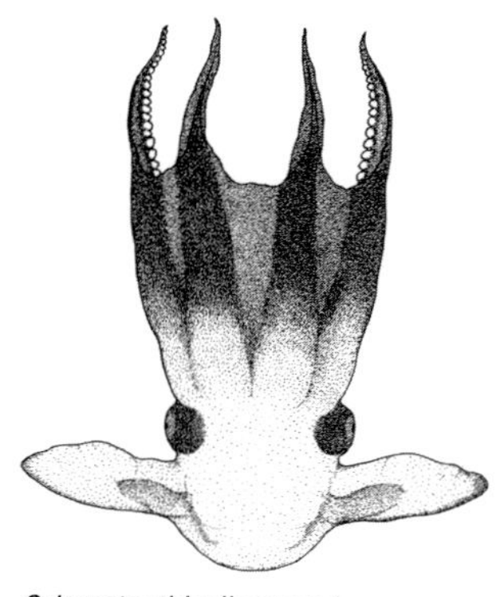

Grimpoteuthis discoveryi

IAN RENDALL

As scientists continue to probe the vast darkness of the Earth's deep-sea trenches, new and strange creatures will continue to be discovered. New species *Grimpoteuthis discoveryi* is one of many octopuses that live in the oceans' midnight zone, where the sun never penetrates. Rarely found, the genus *Grimpoteuthis* is commonly known as the Dumbo octopus after the Disney elephant who could fly with his ears. The "ears" of new species *G. discoveryi* are actually fins that the octopus flaps to propel itself through water, with or without its arms—unlike other octopus species, which just use their arms for propulsion. From "ear" to "ear," the fin span is nearly five and a half inches, over half as wide as the octopus's total length of eight inches. *G. discoveryi* was discovered at depths of up to three miles below the Atlantic Ocean's surface.

Original Dumbo "Roll-A-Book" (1939)

DISCOVERED	northeast Atlantic Ocean
SCIENTIFIC NAME	*Grimpoteuthis discoveryi* Collins 2003
ETYMOLOGY	*Grimpoteuthis* is a genus of octopuses that was named in 1932; *discoveryi* after the RRS *Discovery*, the marine research ship that was used to collect specimens.
CLASSIFICATION	Animalia • Mollusca • Cephalopoda • Octopoda • Opisthoteuthidae

NATIONAL OCEANOGRAPHY CENTRE, SOUTHAMPTON, UK

Big Brain Protist

METHINKS THOU DOTH PROTIST TOO MUCH

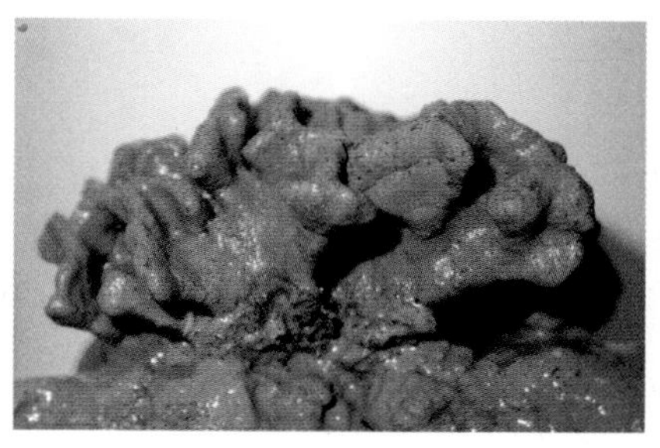

NATIONAL OCEANOGRAPHY CENTRE, SOUTHAMPTON, UK

In 2007, almost three miles beneath the ocean's surface off the coast of Portugal, the unmanned submersible ROV (remotely operated vehicle) *Isis* collected a new species of xenophyophore—a gigantic single-celled organism (protist). With a diameter of up to 4.7 inches, *Reticulammina cerebreformis*'s single cell is *huge* when compared to the multicelled animals that appear in our next chapter—such as a jellyfish (0.04 inches), a moth (0.12 inches), a teensy fish (0.35 inches), and even a chameleon that's only 1.2 inches. Other species of protists can reach diameters of almost eight inches. *R. cerebreformis* is distinctively strange for its brainy-looking appearance and for being especially abundant where it was found with population densities reaching twenty-one individuals per square meter. What a bunch of brainiacs.

DISCOVERED	Eastern Atlantic Nazaré Canyon, Portugal
SCIENTIFIC NAME	*Reticulammina cerebreformis* Gooday, da Silva & Pawlowski 2011
ETYMOLOGY	*Reticulammina* is a genus of protists that was named in 1972; *cerebreformis* is from the Latin *cerebrum*, meaning brain, in reference to the protist's brainlike appearance.
CLASSIFICATION	Chromista • Foraminifera • Xenophyophorea • Psamminida • Psamminidae

JUAN CARLOS ZAMORA / REAL JARDÍN BOTÁNICO–CSIC

Little Grooves Earthstar

FUNGI ON A ROLL

Looking more extraterrestrial than homegrown, earthstars are some of the most alien-appearing species in a kingdom—the fungi—that has more than its fair share of weird. Earthstars don't even *act* like other mushrooms. In fact, they're among the few fungi that can actually move. When the weather is dry, the earthstar's petal-rays curl around its center to protect its round "spore sack," which looks more like its cousin, the familiar puffball. As the earthstar encloses, the spore sack lowers itself to the ground. In its dry state, the earthstar may sometimes detach itself from the soil and start tumbling around. When the weather turns wet, the petal-rays of the earthstars reopen as the spore sack increases in height and disperses spores into the air. Several new earthstar species were discovered this decade including the beautiful lavender *Geastrum episcopale* (Argentina, 2009) and our pick of the weird, the Spanish earthstar *G. parvistriatum*.

JUAN CARLOS ZAMORA / REAL JARDÍN BOTÁNICO-CSIC

Lycoperdon hyemale ("puffballs") Plate 72 from *Histoire des Champignons de la France* by Pierre Bulliard (c. 1780–1791).

COURTESY OF VINTAGE PRINTABLE

DISCOVERED	Spain
SCIENTIFIC NAME	*Geastrum parvistriatum* Zamora & Calonge 2007
ETYMOLOGY	*Geastrum* is a genus of fungi named in 1801; *parvistriatum* is from the Latin words *parvus* for small and *striatum* for striations or "narrow grooves."
CLASSIFICATION	Fungi • Basidiomycota • Agaricomycetes • Geastrales • Geastraceae

NIKOLAJ SCHARFF

Long-Neck Assassin Spider

FANTASTICALLY FEROCIOUS FANGS

HANNAH M. WOOD

If you are a spider, be afraid of assassin spiders. Be very afraid. Less than one-eighth of an inch long, assassin spiders are—millimeter for millimeter—among the most ferocious predators on Earth. Their elongated "necks" and weird fangs make assassin spiders among the most unusual arachnids that are as deadly to other spiders as they are grotesque. Unwary prey are stabbed by venomous jaws ten times longer than those of comparably sized spiders. These absurdly proportioned fanglike appendages are made even weirder by their position atop an equally peculiarly elongated "neck." Spiders, of course, don't have necks. In fact, their head and thorax are fused into a single body region known as the cephalothorax that, in this case, is freakishly and vertically stretched. This latest assassin spider species is found nowhere on Earth except Madagascar, like so many other unique plants and animals of that island nation.

DISCOVERED	Bay of Baly National Park in Mahajanga Province, Madagascar
SCIENTIFIC NAME	*Eriauchenius lavatenda* Wood 2008
ETYMOLOGY	*Eriauchenius* is a genus of assassin spiders named in 1881; *lavatenda* after the Malagasy word meaning "long neck."
CLASSIFICATION	Animalia • Arthropoda • Arachnida • Araneae • Archaeidae

Sahyadri Nose Frog

GREAT GLOBS OF GHATS!

Like an alien species from a 1950s science fiction film, *Nasikabatrachus sahyadrensis* is such an unusual-looking species that it's hard to believe, even if you see it. Measuring approximately five and a half inches from its strange little snout to the end of its body (the vent), *N. sahyadrensis* was discovered in one of Earth's biodiversity hot spots, the Western Ghats of India. Its discovery eluded taxonomists because the frog lives underground most of the year and only surfaces for a few days in order to breed. According to scientists, this new species is part of an ancient lineage that is about 130 million years old and most closely related to frogs that are unique to the Seychelles. The discovery of *N. sahyadrensis* may provide new evidence of India's significance in the evolution of advanced amphibian species at a time when Earth's landmasses were drifting apart.

DISCOVERED	Western Ghats (Sahyadri) of India
SCIENTIFIC NAME	*Nasikabatrachus sahyadrensis* Biju & Bossuyt 2003
ETYMOLOGY	*Nasikabatrachus* from the Sanskrit word *nasika*, meaning nose and the Greek word *batrachos*, frog; *sahyadrensi's* after Sahyadri, another name for the region where the species was found.
CLASSIFICATION	Animalia • Chordata • Amphibia • Anura • Nasikabatrachidae

TIM-YAM CHAN, NATIONAL TAIWAN OCEAN UNIVERSITY

Ausubel's Mighty Claw Lobster

THE LONG ARM OF THE CLAW

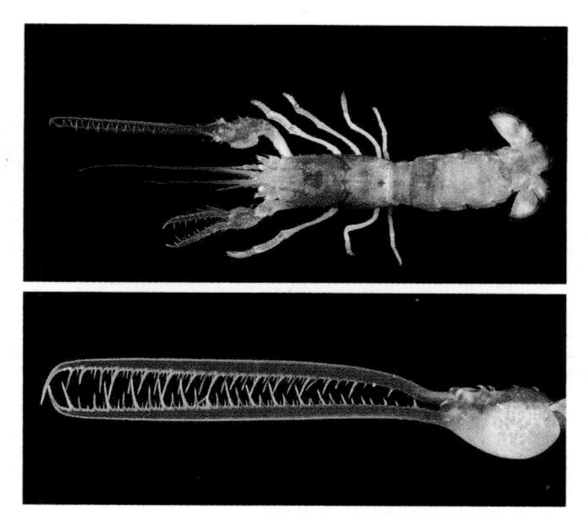

TIN-YAM CHAN, NATIONAL TAIWAN OCEAN UNIVERSITY

Dinochelus ausubeli is a new species that has impressively asymmetrical claws. Discovered during the Aurora 2007 expedition in the deep sea off the Philippines, *D. ausubeli* belongs to a group of lobsters in which the major claw is bulbous at the base with long, narrow, spiny fingers. In a world in which most animals are bilaterally symmetrical, the development of such a hugely unequal pair of claws stands apart. There has not been time to study the biology of the new species in detail, so we can only assume that, like other lobsters, *D. ausubeli* feeds on living and dead prey such as fish, worms, mollusks, and some plant material. Studying lobsters may one day help us extend human life, since lobsters use the enzyme telomerase to repair DNA sequences and show fewer signs of aging than most animals. They do not appear to become weak or slow in old age, and older lobsters are actually more fertile than the youngsters.

DISCOVERED island of Luzon, Philippines

SCIENTIFIC NAME *Dinochelus ausubeli* Ahyong, Chan & Bouchet 2010

ETYMOLOGY *Dinochelus* from the Greek words *dinos*, meaning fearful or terrible, and *chela* for claw; *ausubeli* in honor of Jesse Ausubel, patron of the *Census of Marine Life*, in recognition of his vision and support for marine biodiversity exploration.

CLASSIFICATION Animalia • Arthropoda • Crustacea • Malacostraca • Decapoda • Nephropidae

- A Roosmalen's Hairy Dwarf Porcupine
- B Vanessa's Bamboo
- C Pernambuco Pygmy-Owl
- D Smallest Crustacean
- E Obese Diatom
- F Heckford's Midget Moth
- G Cyprus Mouse
- H Teensiest Chameleon
- I Child of Cypris Tiny Fish
- J CSIRO's Medusa Jelly

Chapter 3

LESS IS MORE, MORE OR LESS

THE SMALLEST NEW SPECIES

As humans—relatively large and self-important animals who believe it's all about *us*—we often forget that we live on a planet where we are totally outnumbered by small-sized organisms. Insects are the most species-rich class of living things, with more than a million named species, and also abundant with 10,000,000,000,000,000,000 (10 quintillion) individuals alive at any given moment. Most of the animals on our planet are small—less than two-fifths of an inch in length. In addition, an estimated 75 percent of living insect species are yet to be discovered and described (about 3 million). Two of the great natural historians of our time, E. O. Wilson and Sir David Attenborough, have pointed out that if humans were to disappear tomorrow, the rest of the world would take little notice and continue more or less as usual. But if the insects were to disappear, most terrestrial and freshwater ecosystems would instantly collapse. Bummer.

But insects are not the smallest, most numerous, or most abundant organisms. Microbial species, too small to be seen by the naked eye, are everywhere. From beneath the ice sheets of Antarctica to boiling hot springs in Yellowstone, from mountaintops to deep-sea vents, the number and diversity of bacteria, archaea, and other microbes, including single-celled protists and fungi, is mind-boggling. But small size does not imply less diversity than in the macroscopic world. Structures of cell walls and the presence or absence of nuclei and various organelles reveal surprising complexity. Some cells actively swim by means of hairlike cilia or whiplike flagella. Physiologically, microbes are astoundingly diverse, coping with extreme conditions of acidity or alkalinity and wide ranges of temperatures and nutrients. Many reproduce at remarkable rates, some doubling their population every twenty minutes. A recent use of DNA sequencing, called metagenomics, has made it possible to do broad environmental sampling of this unseen biodiversity, which has revealed unexpected levels of diversity in both terrestrial and aquatic systems. One recent calculation suggested there may be 20 million microbe species in the oceans alone, with comparable numbers on land.

Life is so diverse that it is difficult to generalize, particularly for groups with large numbers of species. Consider, for example, the range of size among beetles. Some species of featherwing beetles are microscopically tiny at 250 micrometers in length (0.01 inches). These minute insects are smaller than some single-celled species such as *Gromia sphaerica*—an amoeba that is more than an inch in diameter. The longest long-horned beetles are nearly eight inches and the Goliath beetle can weigh three and a half ounces, more than a small rodent. Forty-two of the smallest flowering plants could be lined up end to end within the space of one inch while the largest flower is about the size of a Volkswagen Beetle. A tiny fish reported in this chapter makes the point even more dramatically: at less than three-tenths of an inch (9 millimeters), it is the smallest adult vertebrate animal known. In contrast, the largest vertebrate, the blue whale, measures 110 feet (33,580 millimeters) and is the heaviest animal known to have ever existed—including the great dinosaurs.

Because life arose as microscopic single-celled organisms and we now have millions of large multicellular organisms, it would be natural to assume that there has been a trend from small to large and from simple to complex. But, once again, species are so diverse that no broad generalizations hold. For many species familiar to us there are prehistoric fossil relatives that were significantly and frighteningly larger, such as gigantic crocodiles (*Crocodylus thorbjarnarsoni*), titanic snakes (*Titanoboa cerrejonensis*), monstrous rabbits (*Nuralagus rex*), and colossal ants the size of hummingbirds (*Titanomyrma lubei*). There are many instances of miniaturization, with examples in the Antarctic, in the deep seas, and on islands. In South America, at least eighty-five freshwater fish species are considered to be miniature. The implications of miniaturization vary from group to group. Sometimes body structures persist but become miniaturized. In other cases, body complexity is simplified. The familiar veins in wings of insects add necessary stiffness in larger-sized flighted species, but in tiny wasps, a single vein across the leading edge of the wing is sufficient for flight to take place. Over millions of years, species in one group may evolve in opposite directions to become both larger *and* smaller, or, more *and* less complex as driven by different selection pressures of nature.

COURTESY OF VINTAGE PRINTABLE

As you would probably expect, this chapter presents some of Earth's microscopic, teensiest new species. Some may be new to you, such as a diatom, and some may be as common as a mouse. But you'll also find the smallest species in groups that may be better known for being a lot bigger.

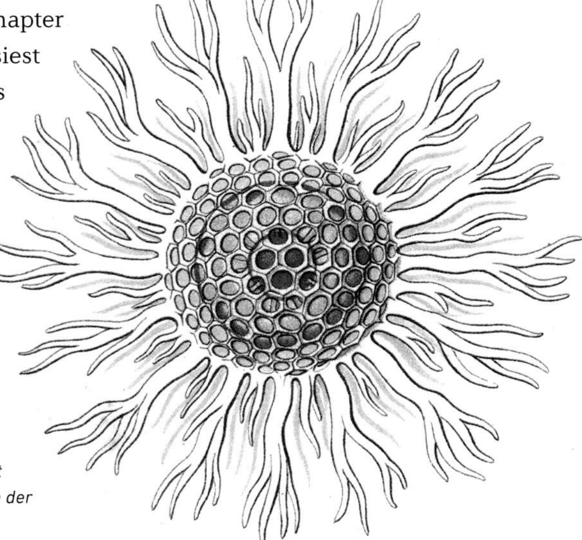

Protozoa, Ernst Haeckel's *Art Forms of Nature, Kunstformen der Natur* (1904).

MARC G. M. VAN ROOSMALEN

Roosmalen's Hairy Dwarf Porcupine

BBQ *PORC ESPIN*

MARC G. M. VAN ROOSMALEN

If you've ever owned a big dumb dog that couldn't resist making friends with the porcupine in the woods, then you know about barbed quills (and visits to the veterinarian). The French also know about porcupines and, in fact, the common name "porcupine" comes from the French words *porc espin*, meaning "spined pig." One of the recently discovered dwarf porcupines, *Coendou roosmalenorum*, has two other types of hair in addition to the typical defensive quills: soft fur and "bristle quills" that are longer and thinner than typical quills but have no barbs. Discovered in Brazil on the banks of the Rio Madeira, the total head and body length of *C. roosmalenorum* is slightly less than a foot long, with a tail almost as long as its ten-inch body. Most other porcupines have an eight- to ten-inch tail, but their bodies are two to three times longer than this new species.

DISCOVERED	Amazonas, Brazil
SCIENTIFIC NAME	*Coendou roosmalenorum* Voss & da Silva 2001
ETYMOLOGY	*Coendou* is a genus of porcupines named in 1799; *roosmalenorum* is named after Marc and Tomas van Roosmalen, whose Madeira collections included this porcupine specimen.
CLASSIFICATION	Animalia • Chordata • Mammalia • Rodentia • Erethizontidae

EMMET JUDZIEWICZ

Vanessa's Bamboo

PANDA APPETIZER

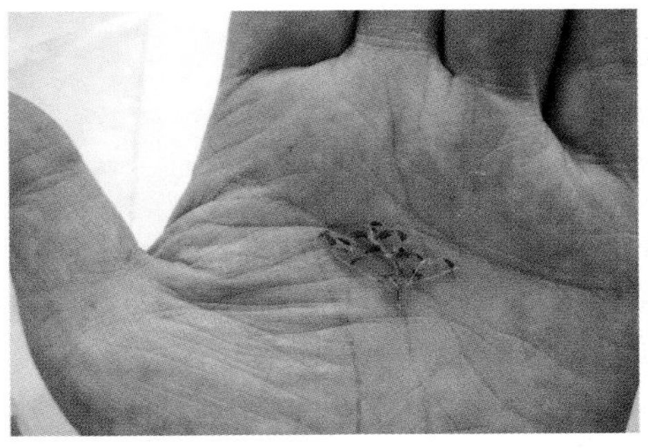

EMMET JUDZIEWICZ

Bamboos are known to be among the most sustainable building materials due to their rapid growth—in fact, some species of bamboo can grow more than three feet in *one day*. Among the one-thousand-plus species in this remarkable grass family is the tiny *Raddiella vanessiae*, which at maturity reaches a height of only three-quarters of an inch—much smaller than its closest bamboo relatives and significantly less than even the daily growth rate of many bamboo species. When the official description of *R. vanessiae* was published in 2007, it was the smallest bamboo known in the world. *R. vanessiae* was recognized as a new species when it was "discovered" at the Smithsonian Institution's National Museum of Natural History, where a specimen had been stored as part of the Biological Diversity of the Guiana Shield Program.

DISCOVERED	French Guiana
SCIENTIFIC NAME	*Raddiella vanessiae* Judziewicz 2007
ETYMOLOGY	The genus *Radiella* was described in 1948; *vanessiae* is named after Vanessa Hequet, who first collected the specimen while studying the ecology of French Guiana in a program funded by the World Wildlife Federation.
CLASSIFICATION	Plantae • Angiosperms • Monocots • Poales • Poaceae

CARL CHRISTIAN TOFTE

Pernambuco Pygmy-Owl

HORTON HEARS A HOOT

Young screech owls from *Annual Report of New York Zoological Society* (1896).

COURTESY OF VINTAGE PRINTABLE

Finding the smallest species of any group is challenging, but finding a small species like *Glaucidium mooreorum*, which is solitary, nearly extinct, and flies at night in a dense forest canopy, is almost impossible. Discovered on a biological reserve in the Atlantic Forest region of northeastern Brazil, this new species of pygmy owl has a two-inch tail and weighs only about 1.8 ounces—in other words, less than the weight of two envelopes. Approximately 12 to 16 percent of known owl species are pygmy owls in the genus *Glaucidium*, and this tiny new species appears to be the smallest owl to be discovered since 1999. Many discoveries of new birds occur when an unrecognized call or song is first heard, not when an unknown bird is first seen. The song of the Pernambuco Pygmy-Owl is so different from its closest pygmy owl relatives that its six-note call helped to determine that it was a new species.

Oscillogram of *Glaucidium mooreorum*'s six-note song.

LUIZ PEDREIRA GONZAGA

DISCOVERED	state of Pernambuco, northeastern Brazil
SCIENTIFIC NAME	*Glaucidium mooreorum* Silva, Coelho & Gonzaga 2002
ETYMOLOGY	*Glaucidium* is a genus of pygmy owls named in 1826; *mooreorum* in honor of Dr. Gordon and Betty Moore, contributors to worldwide and Brazilian biodiversity conservation efforts.
CLASSIFICATION	Animalia • Chordata • Aves • Strigiformes • Strigidae

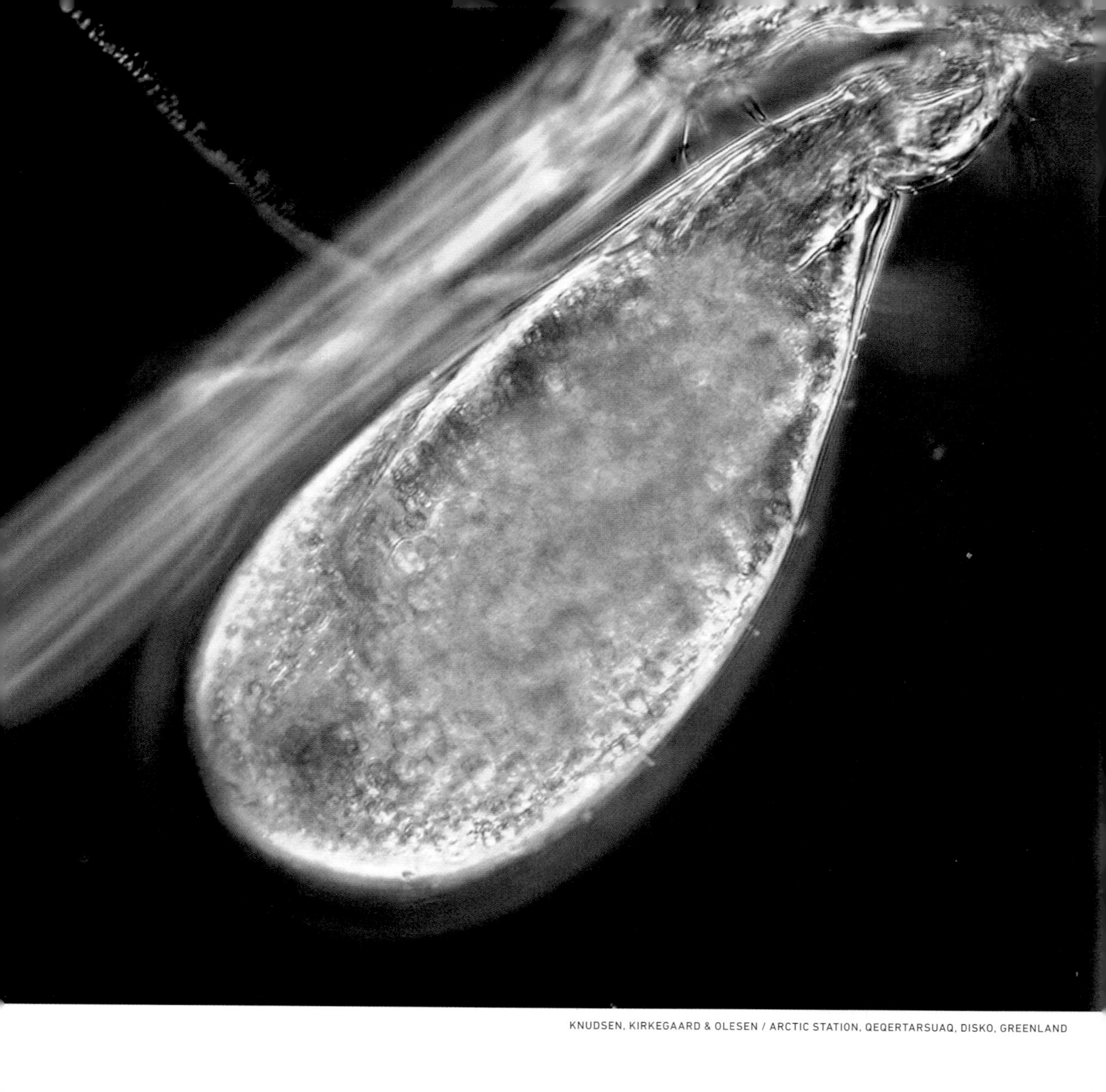

KNUDSEN, KIRKEGAARD & OLESEN / ARCTIC STATION, QEQERTARSUAQ, DISKO, GREENLAND

Smallest Crustacean

DISKO KID

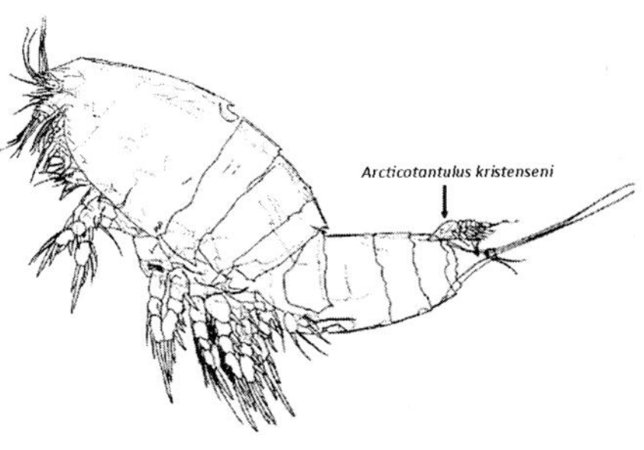

STEEN WILHELM KNUDSEN AND MAJA KIRKEGAARD

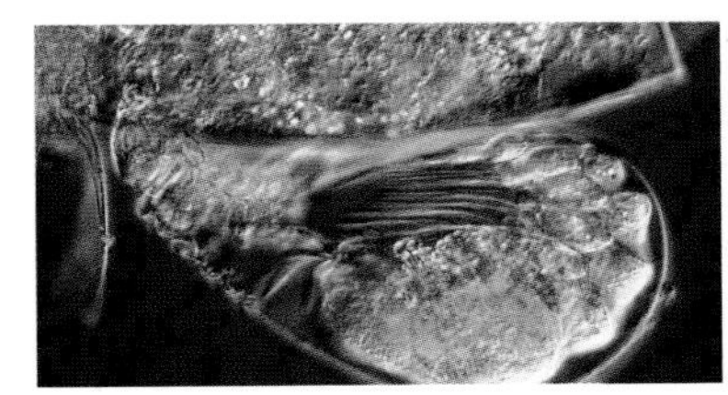

KNUDSEN, KIRKEGAARD & OLESEN/ARCTIC STATION, QEQERTARSUAQ, DISKO, GREENLAND

The world's five oceans contain animals that are the planet's largest, smallest, and every size in between. Among the smallest marine animals are the Tantulocaridans, tiny crustaceans that look like shrimp but are no longer than 0.12 inches. The very smallest of these crustaceans were discovered within the past ten years and can be seen only with a microscope. These teensy animals include the arctic-dwelling *Arcticotantulus pertzovi*, described in 2004 with a total body length of 140 micrometers (0.005 inches), and *A. kristenseni*, described in 2009 with a total body length between 147 and 192 micrometers (about 0.007 inches). At the bottom of the world near the Antarctic are the tiniest of them all—*Tantulacus dieteri* and *T. karolae*, described in 2010 as only 80 micrometers (0.003 inches) in total body length. In fact, *T. dieteri* and *T. karolae* are among the smallest arthropods on Earth. These tiny Tantulocaridans are actually parasites that hitch a ride on their tiny copepod hosts—another group of tiny crustaceans that are only about 0.08 inches long.

DISCOVERED	in two shrimp fishing areas off the west coast of Greenland, in Disko Bay and Disko Fjord
SCIENTIFIC NAME	*Arcticotantulus kristenseni* Knudsen, Kirkegaard & Olesen 2009
ETYMOLOGY	*Arcticotantulus* is a genus of crustaceans named in 2004 by Kornev, Tchesunov & Rybnikov; *kristenseni* named in honor of Professor Reinhardt M. Kristensen, the first to collect the species.
CLASSIFICATION	Animalia • Arthropoda • Crustacea • Maxillopoda • Tantulocarida • Deoterthridae

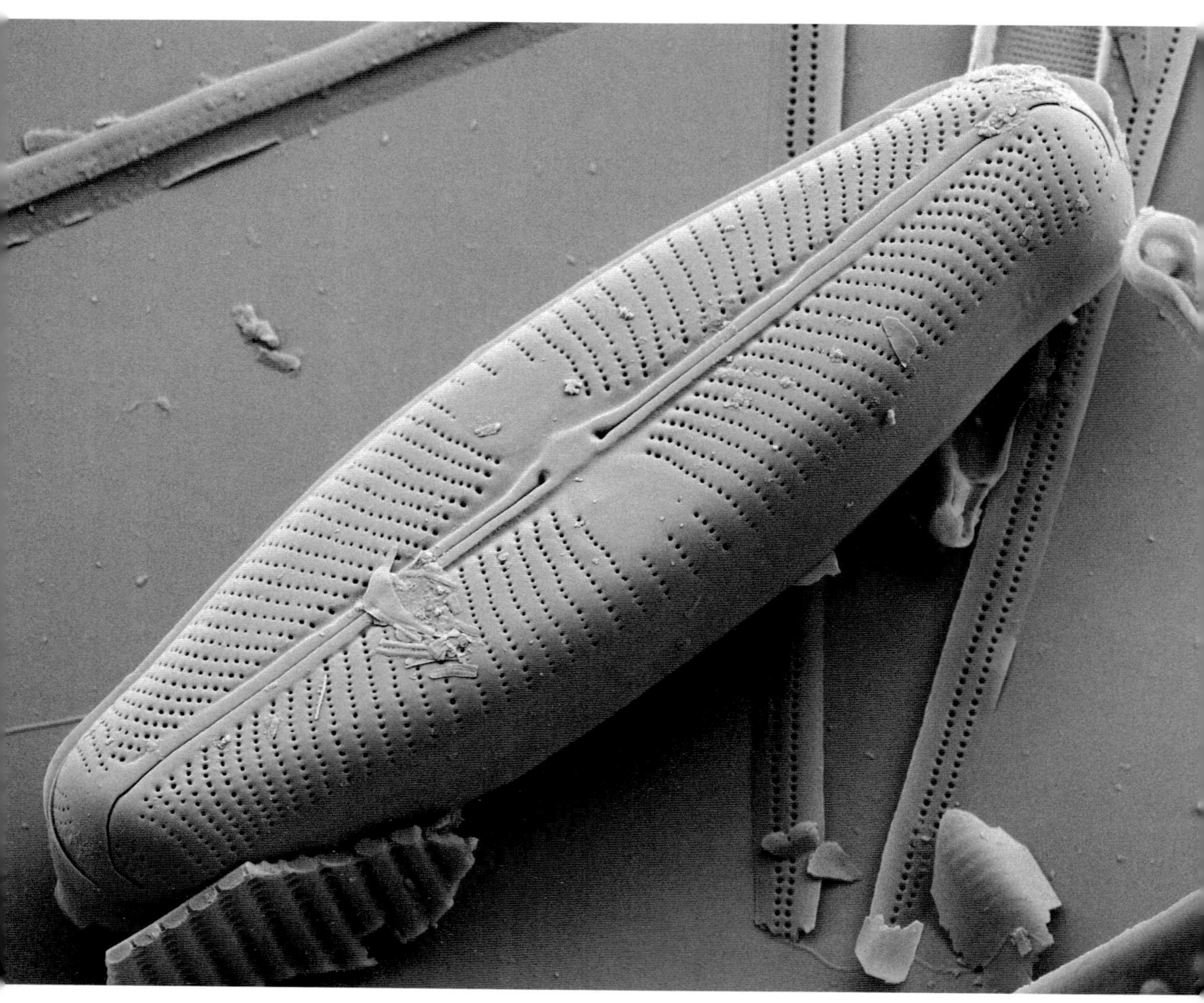

David G. Mann

Obese Diatom

AN ODE TO OXYMORONS

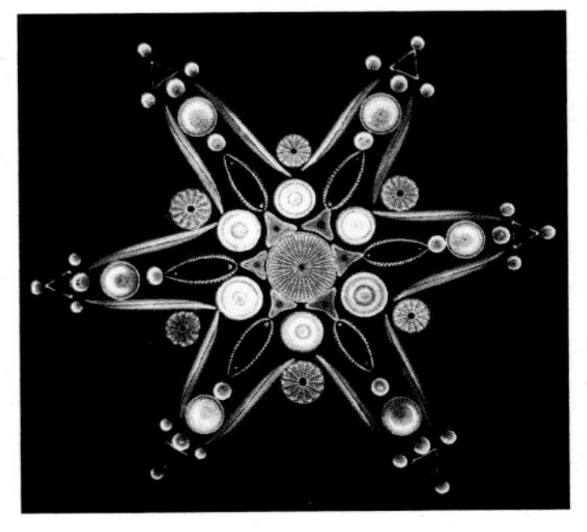

Victorian diatom art, Watson and Sons (c. 1885).

Diatoms are a class of fresh- and saltwater algae that not only provide useful measures of water quality but are also vital for removing greenhouse gases from the Earth's atmosphere. According to University of Washington scientists, diatoms are responsible for 40 percent of the organic carbon produced in the world's oceans each year. Yet these biofuel engineers are so tiny that thirty diatoms side by side would be no wider than a human hair. One of several new diatom species described this century is *Sellaphora obesa*, a beautifully detailed diatom with impressed lines that was found in the accumulated muck of Blackford Pond in Scotland. Despite having a name that means fat, *S. obesa* only measures twenty to forty micrometers (0.0008 to 0.0016 inches), and a thousand *S. obesa* lined up end to end would not add up to an inch. When diatoms were first discovered with the use of early microscopes, Victorian scientists were so enthralled with their beauty and form that they began to arrange them on slides to create the tiniest pieces of art, which are still collected today.

DISCOVERED	Edinburgh, Scotland
SCIENTIFIC NAME	*Sellaphora obesa* Mann & Bayer 2004
ETYMOLOGY	*Sellaphora* is a genus of diatoms named in 1902; *obesa* is the Latin word for fat or plump.
CLASSIFICATION	Chromista • Ochrophyta • Bacillariophyceae • Naviculales • Sellaphoraceae

VAN NIEUKERKEN, ET AL., *ZOOKEYS* 32, CREATIVE COMMONS ATTRIBUTE LICENSE

Heckford's Midget Moth

A MINER DISCOVERY FROM ENGLAND

Pygmy or midget moths of the family Nepticulidae include the smallest of all moths and butterflies, with wingspans measuring as little as 0.12 inches. Larvae are so small that they live in the "mines" they make *between the tissues* in leaves of host plants. The minute green caterpillars of Heckford's Midget Moth were first discovered in the oak saplings of Hembury Woods in Devon, England. Similar-looking leaf mines—the pattern of tunnels in leaves created by insects as they eat—show up in the fossil record, suggesting that this family of moths has been around for nearly 100 million years. Because the insect fauna of the British Isles is among the most completely known in the world, reports of new species of moths are a rare event. This new species, unique to Britain and found by an amateur naturalist, has a wingspan of about 0.2 inches. *Ectoedemia heckfordi* is also an important reminder that the public plays a vital role in species exploration and discovery.

VAN NIEUKERKEN, ET.AL., *ZOOKEYS* 32, CREATIVE COMMONS ATTRIBUTE LICENSE

DISCOVERED	England
SCIENTIFIC NAME	*Ectoedemia heckfordi* Van Nieukerken, Laštůvka & Laštůvka 2009
ETYMOLOGY	*Ectoedemia* is a genus of moths named in 1907; *heckfordi* named in honor of Bob Heckford, the amateur naturalist who first spotted the moth in 2004.
CLASSIFICATION	Animalia • Arthropoda • Insecta • Lepidoptera • Nepticulidae

Cyprus Mouse

MINI MUS

Since 2000, the number of new species discoveries among living mammals has averaged about thirty-six per year, and almost all of them have been in the world's biodiversity hot spots, such as Madagascar, the Amazon jungles, or parts of Asia. Finding a new mammal species in Europe is especially rare. With the exception of a tiny Greek bat identified in 2001 (*Myotis alcathoe*), there hadn't been a new European mammal species discovery for decades until *Mus cypriacus* was found in the vineyards of Cyprus in 2006. This new mouse species is even more remarkable as it is unique to the island and could be considered a "living fossil," since all of the other Mediterranean island rodents became extinct when humans arrived. Sadly, about 15 percent of the mammal species currently living in Europe are threatened by extinction according to the World Conservation Union's (IUCN) 2007 report "The Status and Distribution of European Mammals."

DISCOVERED	Cyprus
SCIENTIFIC NAME	*Mus cypriacus* Cucchi, Orth, Auffray, Renaud, Fabre, Catalan, Hadjisterkotis, Bonhomme & Vigne 2006
ETYMOLOGY	*Mus* is a genus of rodents named by Linnaeus in 1758; *cypriacus* is named after the island of Cyprus, where the species was discovered.
CLASSIFICATION	Animalia • Chordata • Mammalia • Rodentia • Muridae

JÖRN KÖHLER

Teensiest Chameleon

AT YOUR FINGERTIPS

Believed to be the tiniest reptile when it was first discovered in 2007, *Brookesia micra* may also be the tiniest land-dwelling vertebrate in the world. This teensy new species belongs to the leaf chameleons of Madagascar and, according to its discoverers, represents a "striking case of miniaturization" and "an extreme case of island dwarfism." Excluding its tail, *B. micra* males have a maximum length of only 0.63 inches and a total length for either gender of only about 1.2 inches. Like many new species that have been sighted but are not yet officially described by scientists, *B. micra* was given a local or common name and was first called "Nosy Hara" after the tiny island where it was found (at only two sites). Three other tiny chameleon species in the *Brookesia* genus were also discovered during the same 2003–07 expedition to northern Madagascar.

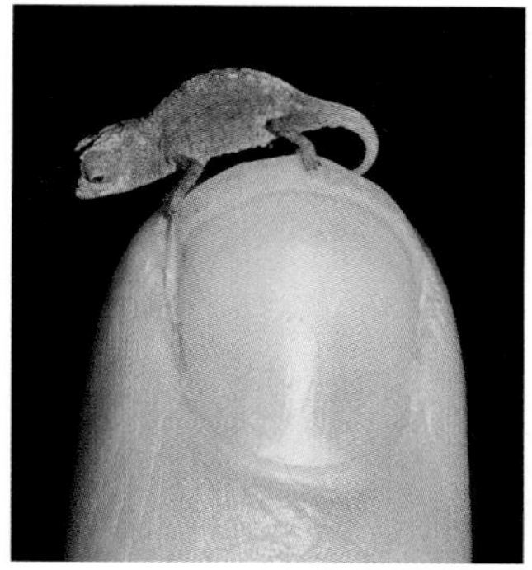

FRANK GLAW

JÖRN KÖHLER

DISCOVERED	Madagascar
SCIENTIFIC NAME	*Brookesia micra* Glaw, Köhler, Townsend & Vences 2012
ETYMOLOGY	*Brookesia* is a genus of tiny Madagascar chameleons named in 1864; *micra* is based on the Greek *mikros* (micros), meaning tiny or small.
CLASSIFICATION	Animalia • Chordata • Reptilia • Squamata • Chamaeleonidae

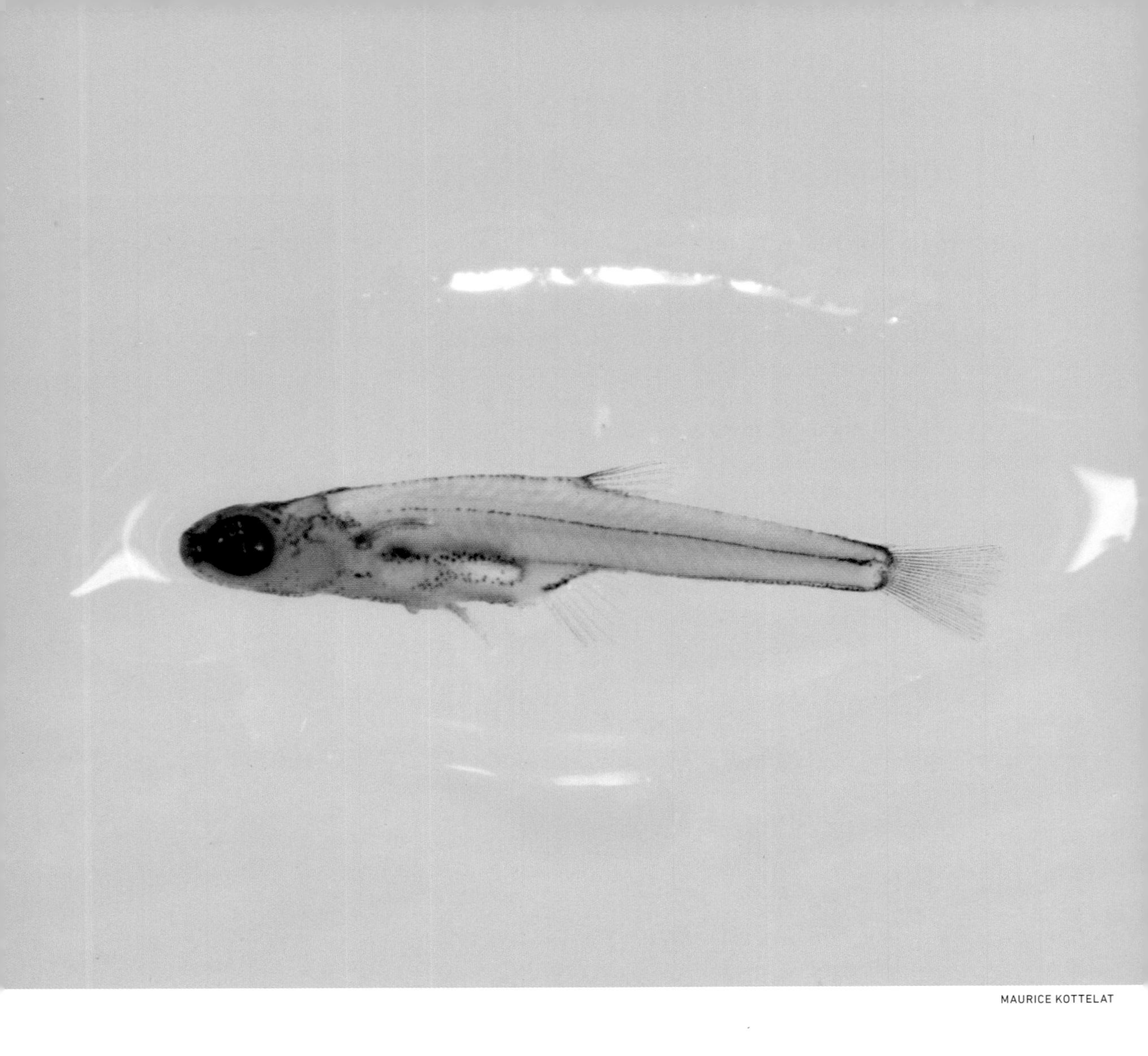

MAURICE KOTTELAT

Child of Cypris Tiny Fish

SHOW A LITTLE BACKBONE

The record for the smallest vertebrate animal was broken in 2004 when an exceedingly small fish, *Schindleria brevipinguis*, was discovered in Australia's Great Barrier Reef. Two years later, a Southeast Asian fish, new species *Paedocypris progenetica*, achieved the feat of being even smaller, with an average body length of 0.35 inches. *P. progenetica* is one of a growing list of miniature fishes found in blackwater peat swamps once thought to have few inhabitants. When an animal retains juvenile characteristics into adulthood it is called paedomorphosis—the inspiration for this species' name. Whether this tiny fish retains its world record as the smallest backboned animal remains to be seen. However, soon after the discovery of this tiny fish was announced, a choreographer from the Netherlands was so inspired he created a ballet and named it *Paedocypris Progenetica* (*Small Fish*).

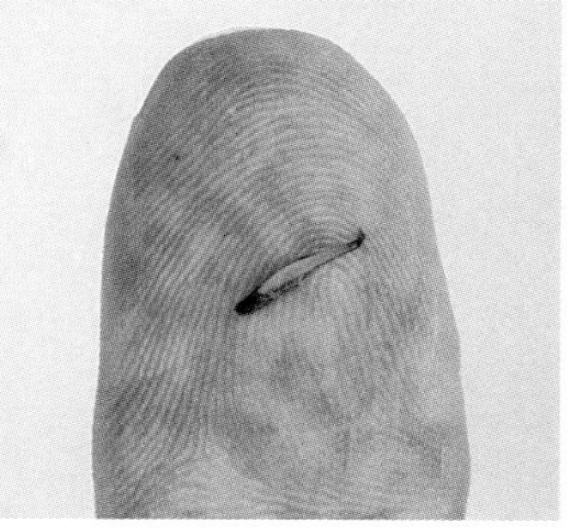

MAURICE KOTTELAT

DISCOVERED	Sumatra and Bintan Island, Indonesia
SCIENTIFIC NAME	*Paedocypris progenetica* Kottelat, Britz, Hui & Witte 2006
ETYMOLOGY	*Paedocypris* from the Greek words *paideios*, meaning children, and *cypris* for Venus (and the common suffix used for fish species in the family Cyprinidae); *progenetica* refers to progenesis and reflects the species' accelerated sexual maturity.
CLASSIFICATION	Animalia • Chordata • Actinopterygii • Cypriniformes • Cyprinidae

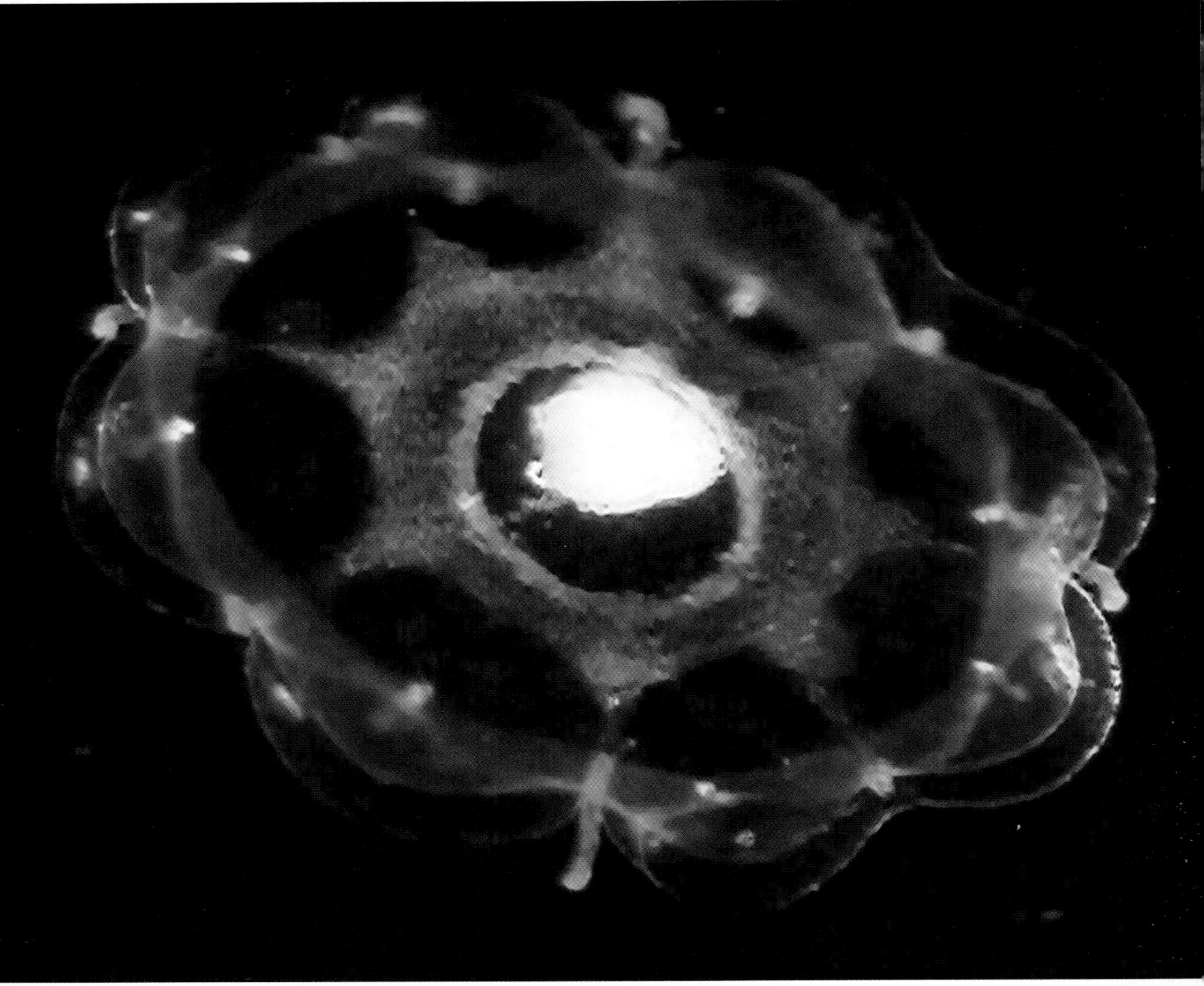

LISA-ANN GERSHWIN

CSIRO's Medusa Jelly

CLOSE ENCOUNTERS OF THE TASMANIAN KIND

When you have to use a glass eyedropper to collect a new animal species, then you *know* it's minuscule. Discovered in 2002 on a harbor wharf in Tasmania, *Csiromedusa medeopolis* is not only tiny, it's incredibly unusual. Naming most new species requires the scientist to create just the second part of its scientific name—the species epithet. This is because the new species can be placed in a genus that may have been described years ago. However, *C. medeopolis* also needed a new genus and a new family to be named—Csiromedusidae—because its body structure was unlike the features of any other jellyfish ever found. Shaped like a flying saucer, *C. medeopolis* has two whorls of tentacles instead of one and its bevel-edged, flattened body contains structures that look like the Manhattan skyline. This tiny new jelly is about 0.04 inches in diameter and was captured in a net with a 500-*micrometer* mesh.

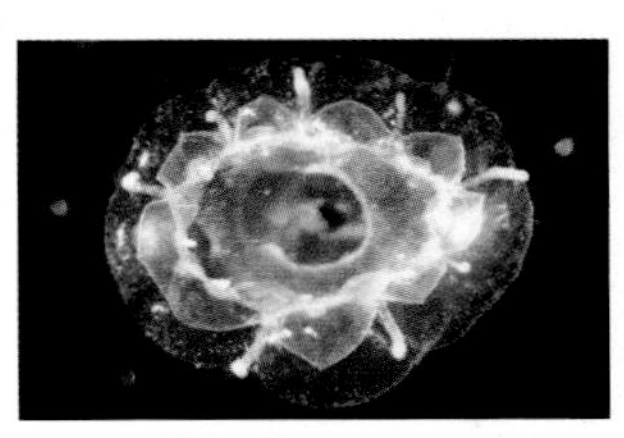

LISA-ANN GERSHWIN

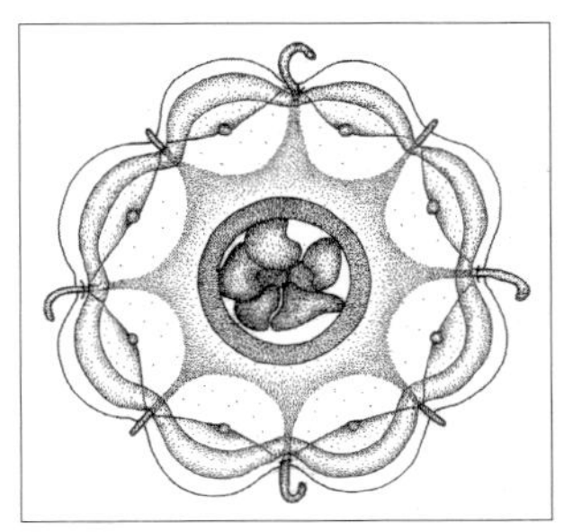

WOLFGANG ZEIDLER

DISCOVERED Hobart, Tasmania

SCIENTIFIC NAME *Csiromedusa medeopolis* Gershwin & Zeidler 2010

ETYMOLOGY *Csiromedusa* is a combination of the acronym CSIRO (Commonwealth Scientific and Industrial Research Organization) and Medusa, the mythological Greek creature who had living snakes for hair; *medeopolis* is from the Greek words *mede* for gonads and *opolis*, meaning city.

CLASSIFICATION Animalia • Cnidaria • Hydrozoa • Narcomedusae • Csiromedusidae

- A Golden V Kelp
- B Big Red Jelly
- C Raptor Fairy Shrimp
- D Solórzano's Velvet Worm
- E Udzungwa Grey-Faced Sengi
- F Huntsman Spider
- G Chan's Mega-Stick
- H Sierra Madre Monitor Lizard
- I Sir Raffles's Showy Flower
- J Idip's Starfish

Chapter 4

BIG DEALS

THE LARGEST NEW SPECIES

Imagine mesmerized crowds lining the streets of ancient Rome when Julius Caesar returned from Africa in 46 BC with the first giraffe ever seen in Europe (*Giraffa camelopardalis*), or a time when throngs gathered on a dock in Lisbon to catch a glimpse of an Indian rhino (*Rhinoceros unicornis*) as it was unloaded from a sailing ship in 1515. Picture Lewis and Clark's astonishment as they first set eyes on giant sequoia trees (*Sequoiadendron giganteum*) on the western coast of North America in 1804. Such jaw-dropping discoveries of unexpectedly large plants and animals continue today—perhaps with less fanfare and drama, but with no less wonder. As we explore the more than 80 percent of plant and animal species unknown to science, records for the largest species of various groups are routinely broken. Largest doesn't necessarily mean huge—it can simply be the longest, heaviest, thickest, or widest within the context of a particular genus or family.

People love to see records broken. And nature does not disappoint, yielding record after record for those with the ambition to search. For instance, the world's largest living insect, measured by body length but not by weight, is a new species of walking stick (*Phobaeticus chani*) from Borneo described in 2008. At twenty-two inches long, it had to be mounted diagonally to fit in a specimen drawer at the Natural History Museum in

London. A species of spider (*Caerostris darwini*) discovered in Madagascar in 2010 weaves the largest webs ever seen, with lengths of up to eighty-two feet, using silk that, ounce for ounce, is ten times tougher than the material used to manufacture bulletproof vests. If we have only just discovered the longest insect and largest spiderweb, what else do we *not* know about the phylum Arthropoda?

Many new size records involve species that lived millions of years ago. Recent examples include a fossil ten-pound "devil frog" (*Beelzebufo*), a one-ton rodent (*Josephoartigasia monesi*), and a terror bird (*Kelenken guillermoi*) that would have towered over a human. Based on its skull and hind leg bone, this huge bird stood ten feet tall, weighed about five hundred pounds, and ran thirty-five miles per hour. In other cases, the record of the largest has been shattered by the discovery of an even larger living species.

How is it possible that large living species have escaped detection by scientists? In some cases, an isolated habitat has only recently been reached. In others, we have only

COURTESY OF VINTAGE PRINTABLE

just invented techniques that make it possible to collect in the places where they live. With improved submersibles and diving gear, we have lengthened our reach into the depths of coral reefs and the sea, and new methods for sampling forest canopies have stretched our reach hundreds of feet above the ground. More often, it is a matter of having the right scientist in the right place. There are so many species that special knowledge is required to definitively tell many species apart. It is frequently the case that when "known" species are critically reviewed by an expert, it is discovered that previously unrecognized species have been there all the time in plain sight. We can be certain of one thing. With 10 million species waiting to be discovered and some unknown number of fossils to be unearthed, records of the largest will continue to be broken as the biosphere is explored.

COURTESY OF VINTAGE PRINTABLE

SHAWN HARPER

Golden V Kelp

KELP WANTED

Volcano alerts and advisories can be common in the Aleutian Islands, where the uncommonly large *Aureophycus aleuticus* (Golden V kelp) was first spotted in the remarkably clear waters off Kagamil Island, Alaska. Kagamil Island is unusual for its high level of thermal activity; gases, steam, and smoke stacks can be seen along the shoreline and the rotten egg scent of sulfur perfumes the air. In this environment, the giant new kelp species *A. aleuticus* was discovered and collected in 2006. Unusually large for kelp, the blades (long leaves) of *A. aleuticus* can be over six feet long and almost two feet wide—that's one *big* sushi roll.

SHAWN HARPER

DISCOVERED	Kagamil Island, Aleutian Islands, Alaska, USA
SCIENTIFIC NAME	*Aureophycus aleuticus* Kawai, Hanyuda, Lindeberg & Lindstrom 2008
ETYMOLOGY	*Aureophycus* from the Latin word *aurum* for gold and the Greek root *phyk* for seaweed, reflecting the kelp's yellow-gold color; *aleuticus* referring to the Aleutian Islands, where the species was discovered.
CLASSIFICATION	Chromalveolata • Ochrophyta • Laminariales • Phaeophyceae

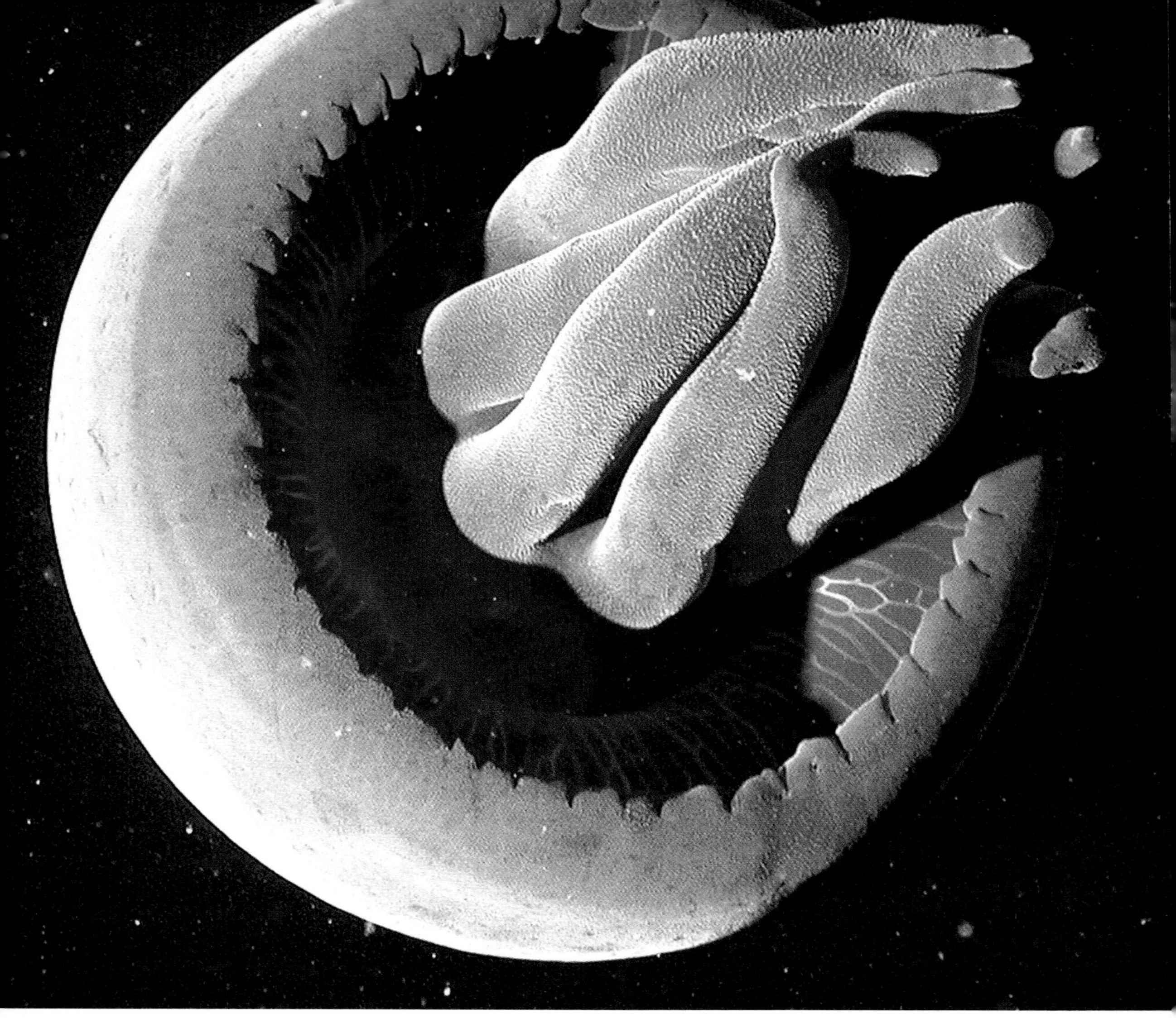

Big Red Jelly

WHAT THE JELL?

Deep in the cold, dark waters of the Pacific Ocean lives *Tiburonia granrojo* (Big Red), one of the world's largest sea jellies in the true jellyfish class Scyphozoa. Not only can this strange new species be found almost a mile below the ocean's surface, its stumpy arms can vary in number between four and seven. Interestingly, *T. granrojo* lacks the tentacles that form the net of appendages that most jellyfish use to catch prey. Big Red lives up to its name—its massive, cannonball-shaped body is blood-red and can reach a diameter of almost three feet. This amazing jelly was discovered by researchers at the Monterey Bay Aquarium Research Institute (MBARI), who used robotic submersibles to collect video and samples of the animal.

© 2002 MBARI

DISCOVERED	Observations near Hawaii and in the eastern and western Pacific, including Gumdrop Seamount, Pioneer Seamount, Taney Seamount, and Davidson Seamount
SCIENTIFIC NAME	*Tiburonia granrojo* Matsumoto, Raskoff & Lindsay 2003
ETYMOLOGY	*Tiburnonia* after the unmanned submersible ROV *Tiburon*, which captured this jelly on video; *granrojo* from the Spanish words *gran*, meaning big, and *rojo*, meaning red.
CLASSIFICATION	Animalia • Cnidaria • Scyphozoa • Semaeostomeae • Ulmaridae

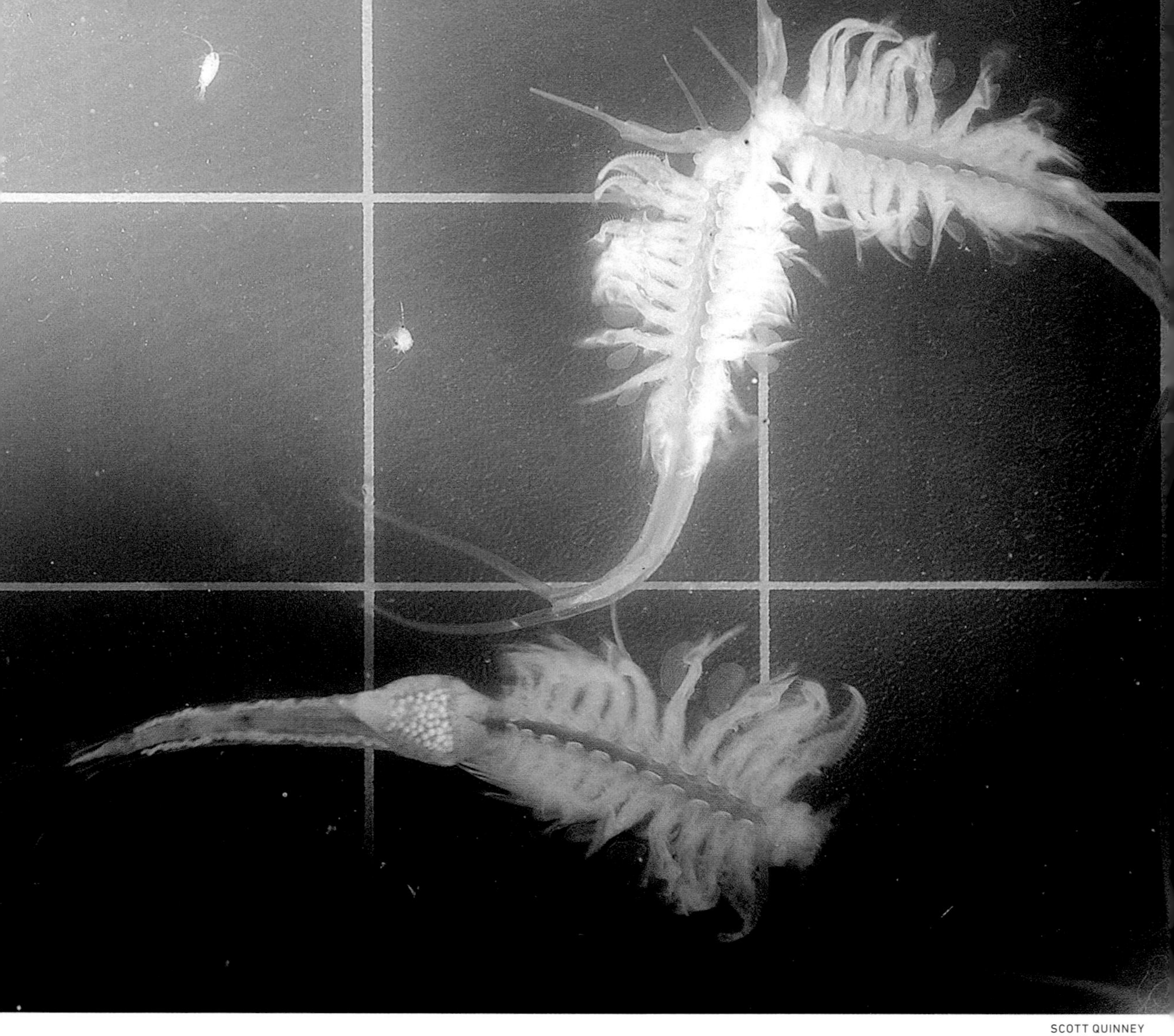

SCOTT QUINNEY

Raptor Fairy Shrimp

GRIM FAIRY TAIL

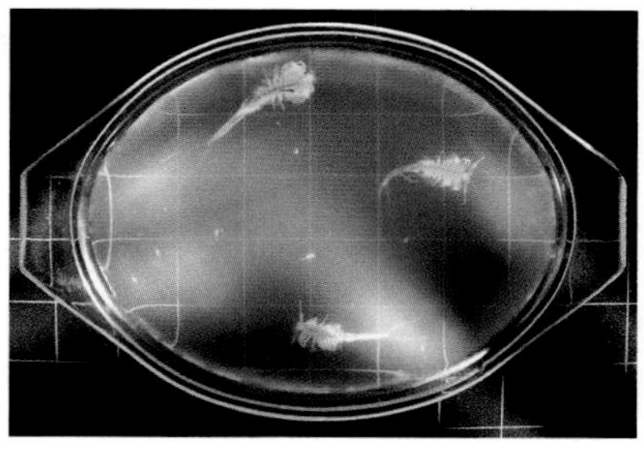

It isn't often that you hear about a species that must eat so frequently that it gets stressed if it hasn't eaten for two and a half hours and starts to die after three hours without food. This new species of fairy shrimp not only exists but also avoids untimely death by carrying its prey around for later snacking. Discovered in freshwater lakes in southern Idaho, *Branchinecta raptor* is a giant and ferocious predatory fairy shrimp that makes most other fairy shrimp look, well, shrimpy. The males are about 2.6 inches and the females are even longer at about 3.2 inches. Both are significantly bigger than their favorite prey, the relatively tiny one-inch *B. mackini*. *B. raptor* often detects its prey by colliding with it, but once that happens, this huge fairy shrimp will attack and chase it. And, like a cat that enjoys toying with its prey, *B. raptor* will repeatedly bite, release, and recapture its favorite meal.

DISCOVERED	Idaho, USA
SCIENTIFIC NAME	*Branchinecta raptor* Rogers, Quinney, Weaver & Olesen 2006
ETYMOLOGY	*Branchinecta* is a genus of crustaceans named in 1869; *raptor* from the Latin *rapere*, meaning to grab and carry off. As the discoverers point out: "Raptor is used in English-speaking countries to refer to birds of prey. This is apt in this case as well, as all currently known localities for this species occur within the Snake River Birds of Prey National Conservation Area."
CLASSIFICATION	Animalia • Arthropoda • Crustacea • Branchiopoda • Anostraca • Branchinectidae

ALEJANDRO SOLÓRZANO, BERNAL MORERA & JULIÁN MONGE

Solórzano's Velvet Worm

LIVING FOSSIL?

About one-third larger than the next largest velvet worm, new species *Peripatus solorzanoi* measures over eight and a half inches. The newborn are even larger than most other Onychophora (velvet worms). This new species belongs to the Peripatidae family, which gives birth to live young rather than laying eggs (upper photo, right). Shortly after birth, *P. solorzanoi* young can spin a glue net to capture prey and defend themselves from predators. Although onychophorans have existed for millions of years and include more than one hundred living species, all are rarely seen and few species are well-known to science. They remain of interest to evolutionary biologists because of their mixture of worm and arthropod characteristics and because they were among the first walking terrestrial animals. For an example of an ancient animal that resembles today's living velvet worms, check out the fossil "walking cactus" in chapter 5.

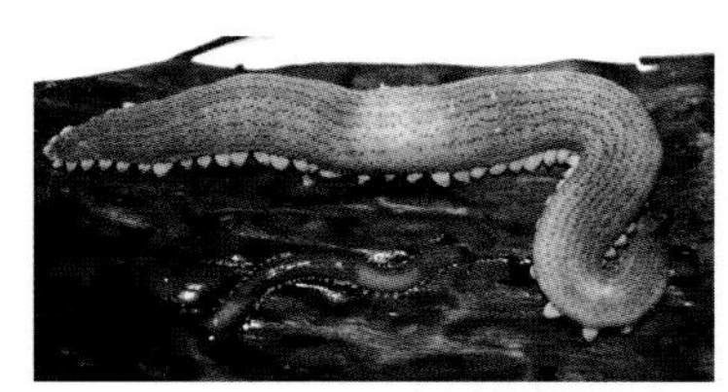

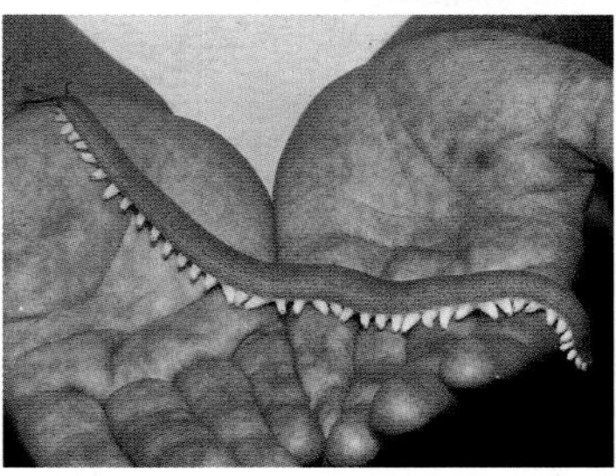

ALEJANDRO SOLÓRZANO, BERNAL MORERA & JULIÁN MONGE

DISCOVERED	Costa Rica
SCIENTIFIC NAME	*Peripatus solorzanoi* Morera-Brenes & Monge-Nájera 2010
ETYMOLOGY	*Peripatus* is a genus of velvet worms named in 1826; *solorzanoi* is named in honor of Alejandro Solórzano, the Costa Rican herpetologist who discovered the species, for his extensive work on the Central American herpetofauna.
CLASSIFICATION	Animalia • Onychophora • Onychophorida • Euonychophora • Peripatidae

FRANCESCO ROVERO / TRENTO MUSEUM

Udzungwa Gray-Faced Sengi

TAMING OF THE SHREW

Elephant shrews—now called sengi—are believed to be more closely related to elephants and aardvarks than to shrews, according to the Afrotheria Specialist Group. New species *Rhynchocyon udzungwensis* is the first sengi to have been discovered in 126 years. It has a single mate throughout its lifetime and can weigh as much as one and a half pounds—about 25 to 50 percent more than other sengi species. Outside of a few families, finding new living land mammals is a rare event, typically occurring in regions that are topographically isolated and difficult to access, such as mountains and dense forests. Discovered in 2006, the gray-faced sengi is an example of a giant new species that can be found in remote regions rich in biodiversity.

FRANCESCO ROVERO/TRENTO MUSEUM

Elephant shrew *Macroscelides typicus*, C. P. Thunberg (1796).

COURTESY OF VINTAGE PRINTABLE

DISCOVERED	Udzungwa Mountains, Iringa Region, Tanzania
SCIENTIFIC NAME	*Rhynchocyon udzungwensis* Rathbun & Rivero 2008
ETYMOLOGY	*Rhynchocyon* is a genus of elephant shrews named in 1847; *udzungwensis* is named after the Udzungwa Mountains in Tanzania, where this species is common.
CLASSIFICATION	Animalia • Chordata • Mammalia • Macroscelidea • Macroscelididae

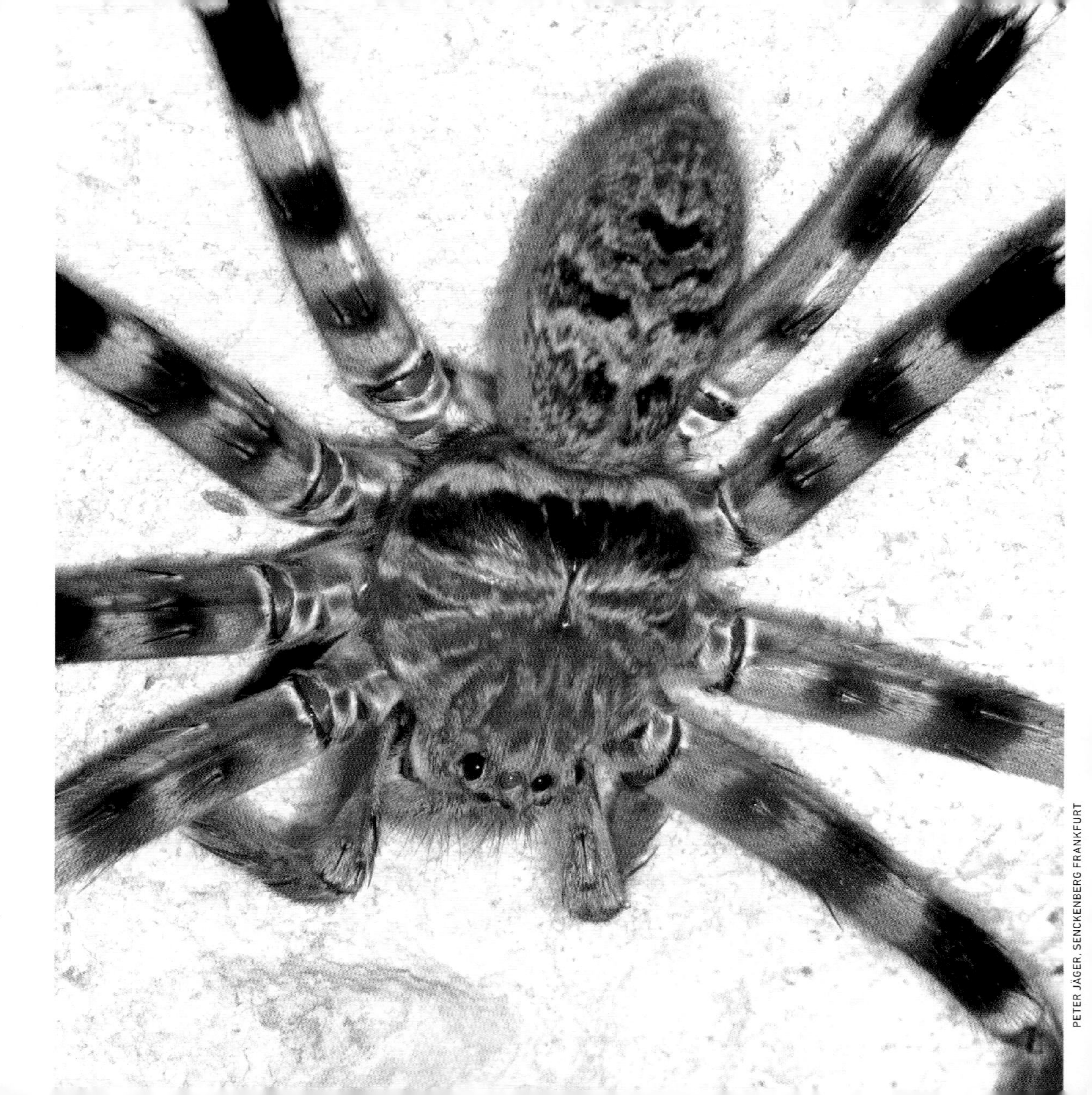

PETER JÄGER, SENCKENBERG FRANKFURT

Huntsman Spider

DISHING OUT ARACHNOPHOBIA

Picture a spider so large that its legs would easily spill over the edges of your dinner plate. In fact, *Heteropoda maxima* has legs that can span up to twelve inches. This colossus is a new species of huntsman spider that is believed to be the largest sparassid, or giant crab spider, in the world. Despite its impressive size, *H. maxima* wasn't discovered and described until 2001, when it was found in a collection of Laotian specimens deposited in the National Museum of Natural History in Paris. Fortunately, *H. maxima* is not deadly to humans, although its size may be fearsome. In addition to living in Laos, members of the Sparassidae spider family can be found in the United States and throughout tropical and subtropical regions of the world, including Australia, China, Japan, and the Philippines.

PETER JÄGER, SENCKENBERG FRANKFURT

DISCOVERED	Cammon (Khammouan), Lao People's Democratic Republic (Laos)
SCIENTIFIC NAME	*Heteropoda maxima* Jaeger 2001
ETYMOLOGY	*Heteropoda* is a genus of spiders named in 1804; *maxima* from the Latin *maximus*, largest.
CLASSIFICATION	Animalia • Arthropoda • Arachnida • Araneae • Sparassidae

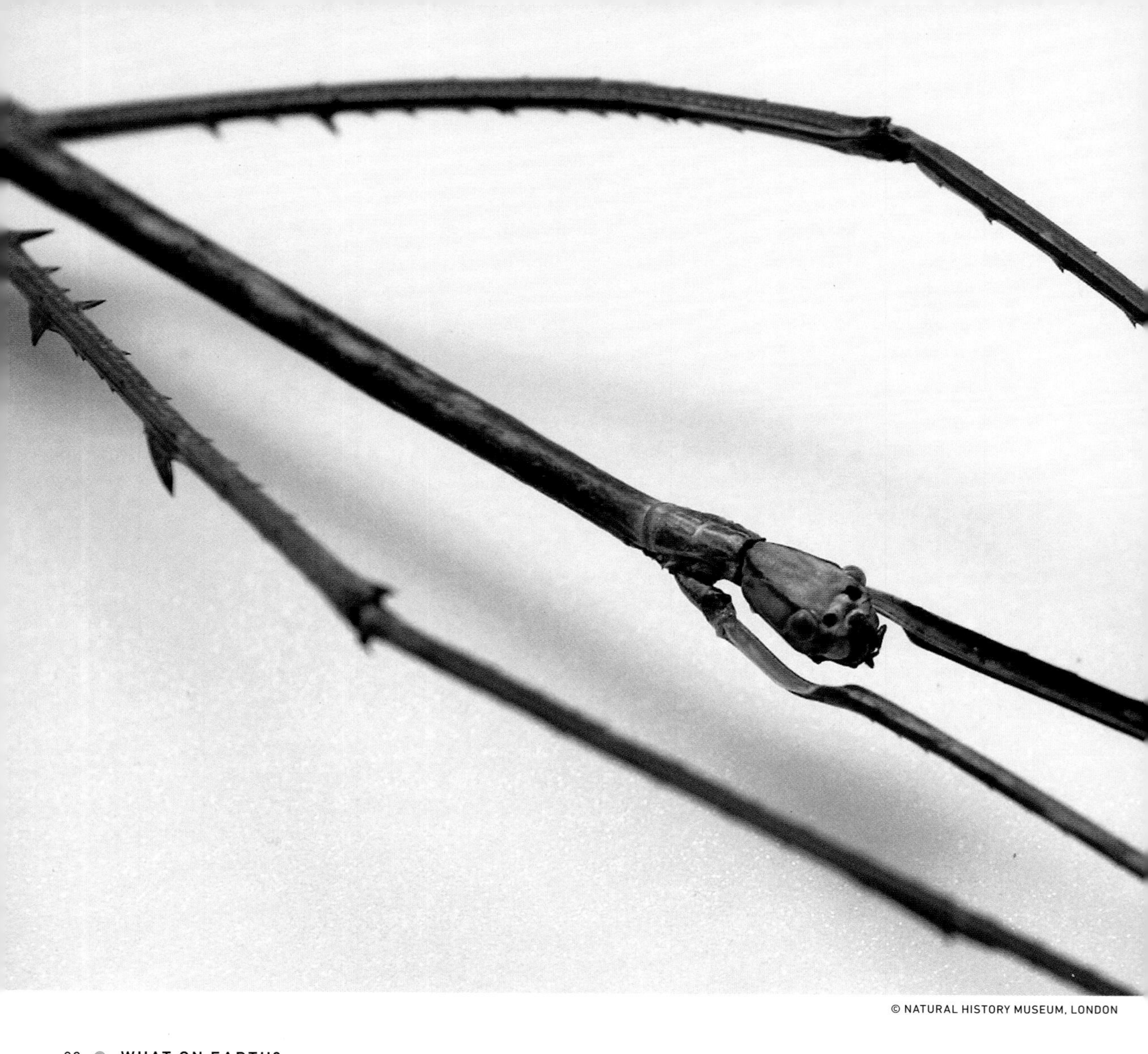

Chan's Mega-Stick

PHANTASTIC PHASMID

Flying eggs and the longest insect body in the world make this walking stick one of the most unique among the largest species to be discovered this century. Originally found by a Borneo farmer, the length of *Phobaeticus chani* (Chan's mega-stick)—including its legs—is nearly as long as an adult human arm (about two feet) while its body length is more than a foot. The leaf-shaped, winglike curvature of *P. chani*'s eggs also makes this new species incredibly interesting. Dropped by the female from high in a tree, the eggs float in the wind or in water, which may allow a distinct evolutionary advantage for species dispersal through protection from hard landings or drowning.

Datuk Chan Chew Lun with his namesake, P. chani.

Egg of *P. chani.*

DISCOVERED	Borneo, Malaysia
SCIENTIFIC NAME	*Phobaeticus chani* Bragg 2008
ETYMOLOGY	*Phobaeticus* is a genus of stick insects named in 1907; *chani* named in honor of C. L. Chan (Datuk Chan Chew Lun) from Kota Kinabalu, Sabah, Malaysia.
CLASSIFICATION	Animalia • Arthropoda • Insecta • Phasmatodea • Phasmatidae

ARVIN DIESMOS

Sierra Madre Monitor Lizard

FRUIT OF THE LUZON

JOSEPH BROWN

Despite years of rumors and the occasional photograph that would come to the attention of biologists, this spectacular fruit-eating and tree-climbing giant monitor lizard, *Varanus bitatawa*, has only recently been described by scientists although it's been known to the local Philippine tribespeople for centuries. The ability of *V. bitatawa* to hide from humans is especially remarkable given its basketball-player size—over six and a half feet in length—and its distinctive bright coloring of yellow spots and stripes. Species of monitor lizards are believed to be highly intelligent. This new species is closely related to Gray's monitor (*V. olivaceus*), which has a similar Philippine forest habitat and is known to be secretive and shy. The combination of street smarts, guarded behavior, and the ability to climb trees has undoubtedly helped this new species remain elusive. Its scientific discovery highlights the importance of identifying the biodiversity of species that live in the Earth's rain forests as well as the need to conserve these habitats.

DISCOVERED	Luzon Island, Philippines
SCIENTIFIC NAME	*Varanus bitatawa* Welton, Siler, Bennett, Diesmos, Duya, Dugay, Rico, van Weerd & Brown 2010
ETYMOLOGY	*Varanus* is a genus of lizards named in 1820; *bitatawa* is from the local common name of the Philippine Agta tribespeople.
CLASSIFICATION	Animalia • Chordata • Reptilia • Squamata • Varanidae

JULIE F. BARCELONA

Sir Raffles's Showy Flower

ESPRIT DE CORPSE

New species *Rafflesia speciosa* is one of the largest *single* flowers in the world, appearing to grow directly on the ground without stems or leaves. Found by members of a Philippine conservation group, this big bud is related to Earth's heaviest and largest individual flower, *R. arnoldii*, which can weigh as much as twenty-two pounds and reach a span of almost forty inches. *R. speciosa* and the more recently found *R. mira* (2006) are considered to be medium-sized *Rafflesia* even though they can be two feet wide. Despite being red or orange and amazingly large, most *Rafflesia* are considered rare and difficult to find in the densely vegetated forests where they grow. The second-largest *Rafflesia*, *R. schadenbergiana*, is so rare that it was presumed to be extinct, since it had not been seen from the time of its discovery in 1882 until 1994. To attract insects for pollination, *Rafflesia* flowers emit an odor that smells like rotting flesh, which earns them the local name of corpse flower.

JULIE F. BARCELONA

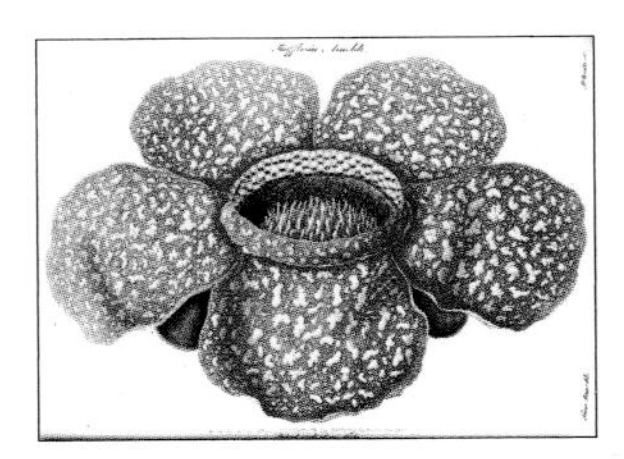

Rafflesia arnoldii historic illustration (1822).

COURTESY OF VINTAGE PRINTABLE

DISCOVERED	Sibalom Natural Park, Panay Island, Philippines
SCIENTIFIC NAME	*Rafflesia speciosa* Barcelona & Fernando 2002
ETYMOLOGY	*Rafflesia* is a genus of flowering plants named in 1818 to honor Sir Thomas Stamford Bingley Raffles, a British statesman who founded Singapore; *speciosa* from the Latin, meaning spectacular, showy, impressive.
CLASSIFICATION	Plantae • Angiosperms • Eudicots • Malpighiales • Rafflesiaceae

CORAL REEF RESEARCH FOUNDATION

Idip's Starfish

A STAR IS BORN

CORAL REEF RESEARCH FOUNDATION

Also called starfish, sea stars can be found in a dazzling array of colors, shapes, and sizes. They can look like armless pincushions or have fifty arms and more. Or they can have just five fat stumpy arms like new species *Astrosarkus idipi*, one of the planet's largest sea stars based on its massive volume. Discovered during the Palau Twilight Zone Expedition of 1997 but not described until 2003, this new species measures almost two feet in diameter, with a fleshy body that can be five inches thick. Colored like a pumpkin with a creamy white mustache, *A. idipi* is especially remarkable because most of its endoskeleton (the skeletal support structure) has been lost during its evolutionary history, making you wonder how it stays in shape.

DISCOVERED	Uchelbeluu (Agulpelu) Reef, Koror, Palau
SCIENTIFIC NAME	*Astrosarkus idipi* Mah 2003
ETYMOLOGY	*Astrosarkus* from the Greek words *aster* for star and *sarkus*, meaning fleshy; *idipi* in honor of David K. Idip, former director of the Bureau of Natural Resources and Development of Palau.
CLASSIFICATION	Animalia • Echinodermata • Asteroidea • Valvatida • Oreasteridae

- A Finney's Bat
- B Aurora Horseshoe Crab
- C Sarmatian Seahorse
- D Sahel Man
- E Levant Octopus
- F Burma Bee
- G Old-Old Mushroom
- H Walking Cactus
- I Half-Shell Turtle
- J Dila's Flower

Chapter 5

SOMETHING OLD, SOMETHING NEW

THE OLDEST NEW SPECIES

In 1868, at the corner of Broad and Sansom Streets in Philadelphia, British sculptor Benjamin Waterhouse Hawkins forever changed the public's relationship with fossils. Under the watchful eye of paleontologist Joseph Leidy, Waterhouse had finished mounting the skeleton of *Hadrosaurus foulkii*, the first reconstructed dinosaur ever put on public display. Humans had finally come face-to-face with a prehistoric animal from the Cretaceous Period and were absolutely captivated.

Just as astronomers study distant stars to gaze into the remote past of the universe, paleontologists study fossils to observe bits of our biosphere from times long gone. Anyone who has visited the Great Serpent Mound (United States), Newgrange (Ireland), Machu Picchu (Peru), or Tassili n'Ajjer (Algeria) will instantly understand the significance of physical contact with prehistoric places and objects and be easily convinced that fossil discoveries are important.

Today the discovery of the world's oldest fossils is evolutionary biology's answer to *Antiques Roadshow*—you can never predict what a paleontologist or fossil hunter will find next. Each time an older fossil is unearthed, scientists are forced to reassess their understanding of evolutionary history and confirm—or challenge—existing ideas. Fossil finds may document a previously unknown geographic distribution or reveal an unimagined body plan that failed to adapt and went extinct millions of years ago. Sometimes a fossil yields a new combination of characteristics that solves a chicken-and-egg riddle: for example, whether flight or biological sonar came first in the evolution toward modern bats. Fossils are part of every nation's prehistoric past and become treasured public heirlooms when they're deposited in museums and scientific collections around the world.

As this chapter's stories tell, fossils come in many forms. In general, hard body parts—such as shells of ancient clams, bones of early mammals, and petrified wood—survive for millions of years. In other instances, we find imprints of leaves sandwiched in layers of sedimentary rocks or discover complete insects trapped in the sap of a tree destined to become amber. Sometimes we find trace fossils that depict the activities of an organism rather than the thing itself—such as dinosaur footprints or insect tunnels in wood.

Fossils also provide glimpses into ancient ecosystems through evidence of co-occurring species and by showing animal and plant responses to environmental changes. Across geologic time is evidence of immigration, emigration, evolution, and extinction preserved in rock strata. The ages of these strata are measured in millions of years across four eras: the Precambrian, the Paleozoic, the Mesozoic, and the Cenozoic. The line between the last period in the Mesozoic (Cretaceous) and the first period of the Cenozoic (Tertiary) marks the famous Cretaceous-Tertiary (K-T) boundary event—the fifth and most recent "mass extinction," when more than half of all living things disappeared, including the large dinosaurs. Theoretical advances since the 1960s allow precise reconstruction of evolutionary history through comparisons of living species, yet fossils hold a special place in the study of biodiversity and continue to provide unique and unpredicted insights.

In this chapter are ten recent fossil discoveries, each of which sets a new benchmark for the oldest specimen of a taxonomic group. There were literally hundreds of candidate fossils to choose from and every one is a remarkable discovery in its own right. In spite of the fact that fossilization was a rare event, it seems that amazing discoveries will continue indefinitely and be constrained only by how diligently we search.

COURTESY OF VINTAGE PRINTABLE

AMERICAN MUSEUM OF NATURAL HISTORY

Finney's Bat

CAN YOU HEAR ME NOW?

Bat wingspan, Malosse de Geoffroy, from *Encyclopédie d'histoire naturelle* (1850).
COURTESY OF VINTAGE PRINTABLE

Wyoming is home to cowboys, rodeos, wide-open spaces, breathtaking national parks, and the spot where the oldest known fossil bats in the world can be found. About 52 million years ago in the Early Eocene, fine layers of mountain lake sediments were being deposited to create the Green River Formation. At the same time, mammals were evolving flight. As they took to the air, the first bats emerged. Discovered in 2003, *Onychonycteris finneyi* is the earliest documented bat found to date. Based on its physical structures, this fossil provides evidence that bats could fly *before* biological sonar—echolocation—had evolved. Such echolocation helps modern bats detect their nighttime whereabouts and locate insects on the wing. The spectacular fossil of this early species clearly shows claws on all of its toes, signifying that this bat most likely could walk on all four feet and hang upside down under tree branches.

DISCOVERED	Fossil Butte Member, Green River Formation, Wyoming, USA
SCIENTIFIC NAME	*Onychonycteris finneyi* Simmons, Seymour, Habersetzer & Gunnell 2008
ETYMOLOGY	From the Greek words *oncho* for clawed and *nycteris* for bat; *finneyi* in honor of Bonnie Finney, who collected the fossil.
CLASSIFICATION	Animalia • Chordata • Mammalia • Chiroptera • Onychonycteridae
PERIOD OF EXISTENCE	Early Eocene, 52 million years ago

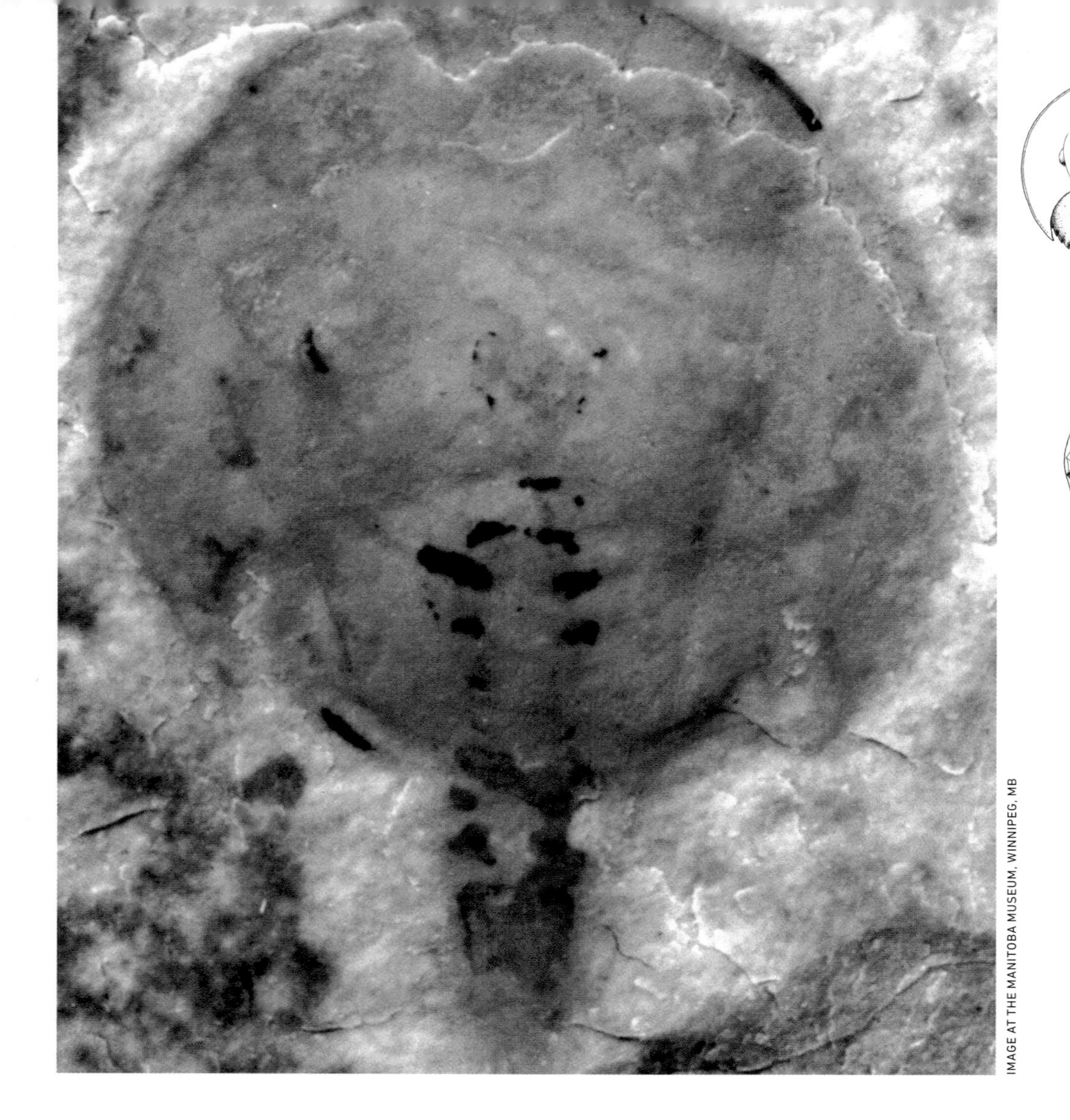

IMAGE AT THE MANITOBA MUSEUM, WINNIPEG, MB

Aurora Horseshoe Crab

A DEAD RINGER

Almost identical in appearance to modern-day horseshoe crabs, *Lunataspis aurora* is a remarkable fossil species discovery that provides evidence of evolutionary stasis—instances when species remain substantially unchanged over millions and millions of years. The discovery of *L. aurora* also shows that the horseshoe crab body plan familiar to us existed far earlier than previously thought. Two of these extraordinary fossil horseshoe crab specimens were discovered during 2005 in Late Ordovician sedimentary deposits in Manitoba, Canada—an area that is especially rich in well-preserved and relatively complete fossils.

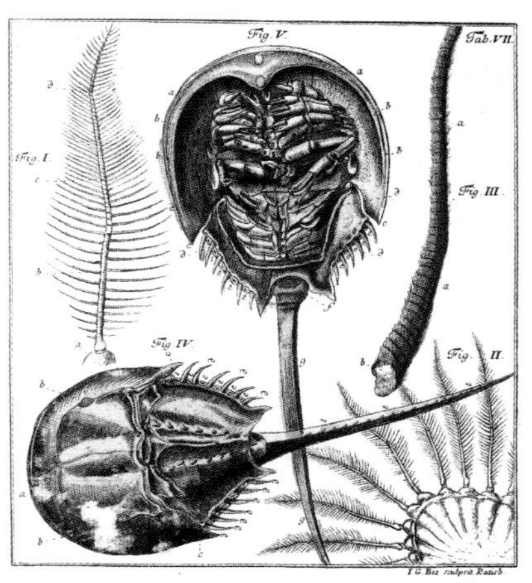

Horseshoe crab, historic illustration (c. 1700s).

COURTESY OF VINTAGE PRINTABLE

DISCOVERED	Airport Cove site, Churchill, Manitoba, Canada
SCIENTIFIC NAME	*Lunataspis aurora* Rudkin, Young & Nowlan 2008
ETYMOLOGY	*Luna* from the Latin, meaning moon or crescent moon, and *aspis* from the Greek, meaning shield; *aurora* from the Latin, meaning dawn and the Roman goddess of dawn.
CLASSIFICATION	Animalia • Arthropoda • Merostomata • Xiphosurida • Limulidae
PERIOD OF EXISTENCE	Late Ordovician, 450 to 460 million years ago

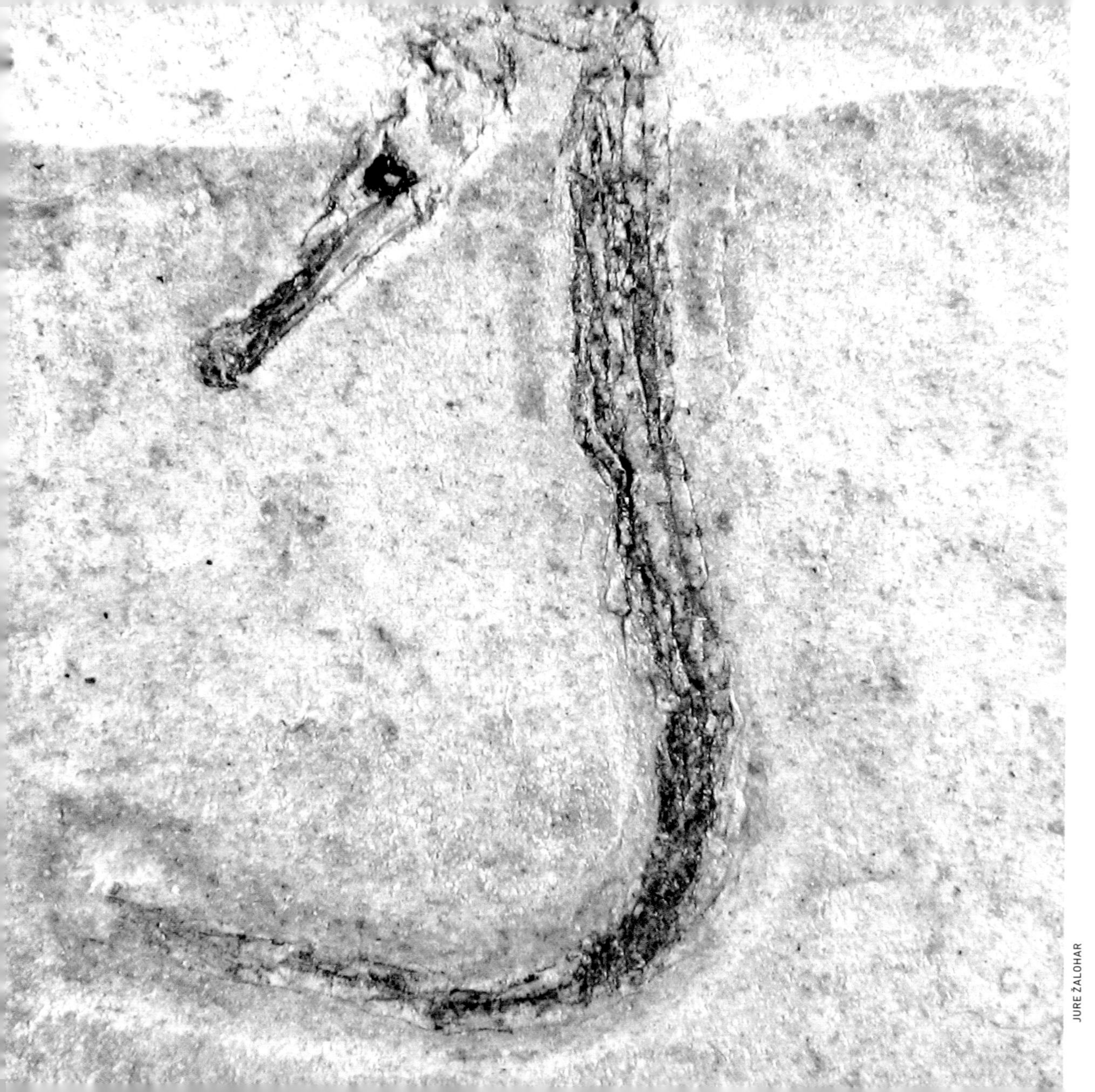
JURE ŽALOHAR

Sarmatian Seahorse

SEAHORSING AROUND

About 15 million years ago, *Hippocampus sarmaticus* lived in the Paratethys Sea—an ancient body of water that extended from the central European Alps to the western boundaries of Asia. Discovered in Slovenia during a 2006 expedition, *H. sarmaticus* and another miniature seahorse species, *H. slovenicus*, are not only the oldest known fossil seahorses by 10 million years but also the only known species of seahorses to be extinct. Tiny in size, they are most closely related to currently living pygmy seahorses and provide evidence that seahorses had a wide geographic distribution during the Miocene Epoch.

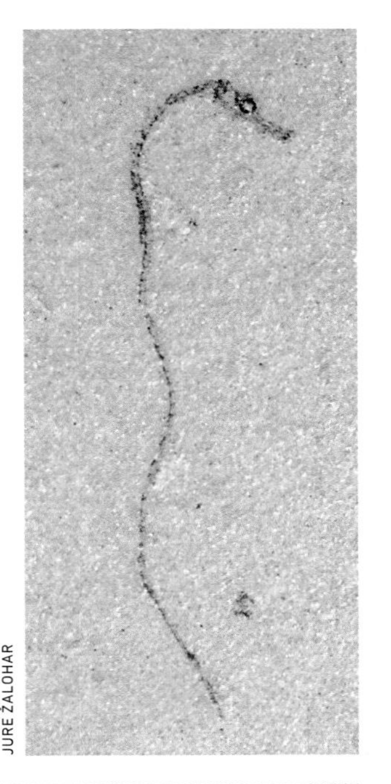

JURE ŽALOHAR

DISCOVERED	Sarmatian deposits in the Tunjice Hills, Slovenia
SCIENTIFIC NAME	*Hippocampus sarmaticus* Žalohar, Hitij & Križnar 2009
ETYMOLOGY	*Hippocampus* is a genus of ray-finned fish named in 1810; *sarmaticus* after the Sarmatian stage of the Middle Miocene and reflecting the geographic area where they were found.
CLASSIFICATION	Animalia • Chordata • Actinopterygii • Syngnathiformes • Syngnathidae
PERIOD OF EXISTENCE	Middle Miocene, Sarmatian, 14 million years ago

Sahel Man

THE TOUMAË SHOW

In 2002, a hominid fossil species, *Sahelanthropus tchadensis* (Toumaë), was found in Africa's "Cradle of Humanity" only one year after *Orrorin tugenensis* (Millennium Man) had been unearthed in the same region. At the time of Millennium Man's discovery, it was believed to be the oldest human ancestor, but Toumaë's discovery had scientists arguing as to which hominid fossil is truly the oldest. Both are about 6 to 7 million years old; both may have walked upright in humanlike fashion; and each potentially represents the evolutionary split between apes and humans-to-be. The age of these fossils seems undisputed—both are significantly older than the 3.2-million-year-old Lucy (*Australopithecus afarensis*), which was discovered in 1974 and, at the time, was the oldest known human ancestor.

DISCOVERED	western Djurab Desert, northern Chad
SCIENTIFIC NAME	*Sahelanthropus tchadensis* Brunet et al. 2002
ETYMOLOGY	*Sahelanthropus* is named after Sahel, the African region where the fossil was found, and the Greek word *anthropos*, meaning human or man; *tchadensis* refers to the country of Chad where the fossil was found.
CLASSIFICATION	Animalia • Chordata • Mammalia • Primates • Hominidae
PERIOD OF EXISTENCE	Upper Miocene, 6 to 7 million years ago

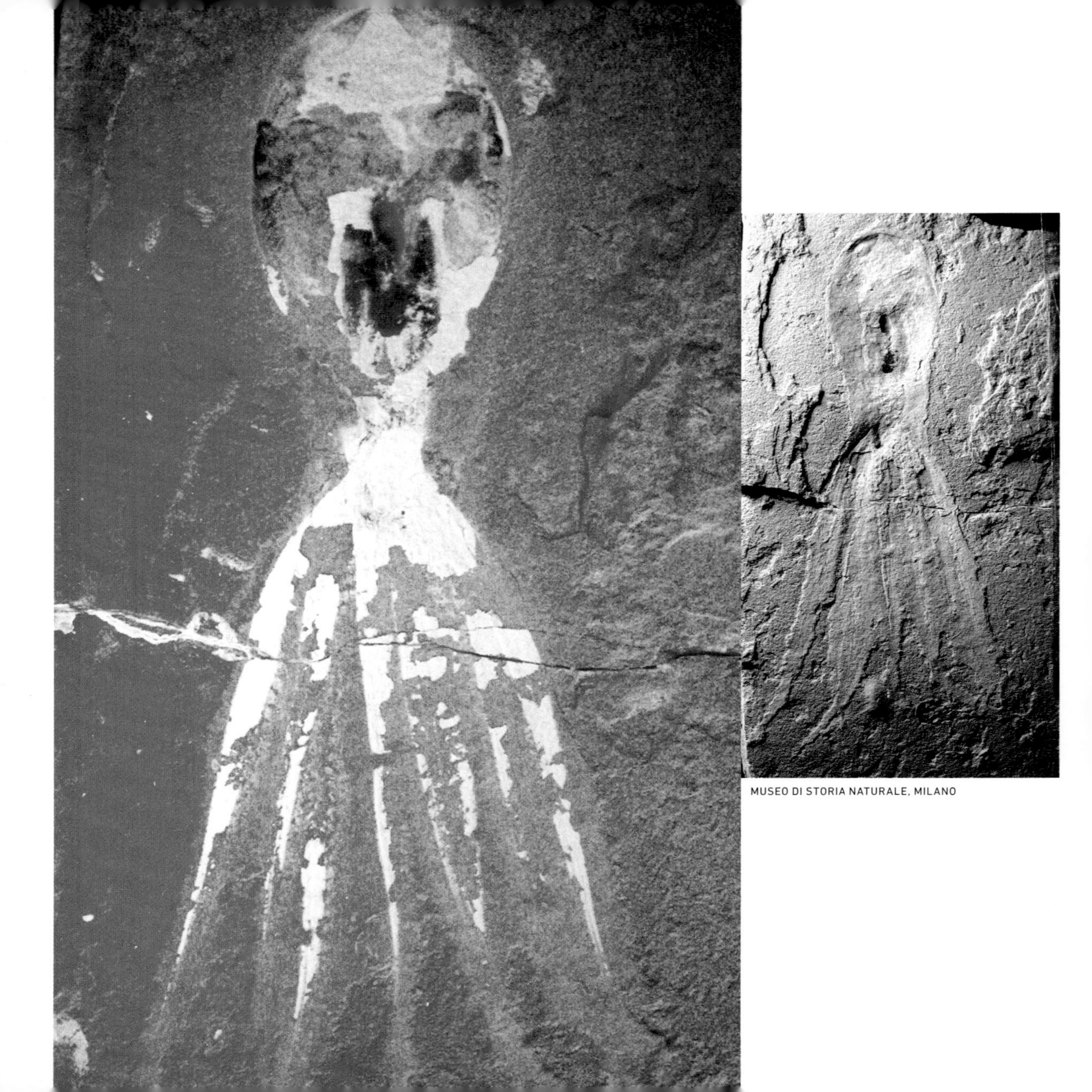

MUSEO DI STORIA NATURALE, MILANO

Levant Octopus

OUTRAGEOUS CRETACEOUS OCTOPOD

According to ancient legends, the Mediterranean Sea was ruled by a Levantine god with many legs and heads who was called Yam by the people of the region now known as the Middle East. Multilegged like Yam and living in the ancient seas, *Keuppia levante* was discovered with two other fossil octopod species new to science, *K. hyperbolaris* and *Styletoctopus annae*. The scientists who discovered these fossils believe them to be the oldest unmistakable examples of Octopoda. The precise age of the limestone rock where these species were found in northwest Lebanon has been debated since 1833. The latest evidence suggests that these species lived about 95 to 180 million years ago, placing them in the Cretaceous or Jurassic Period. Finding these earliest octopods is especially remarkable since soft-bodied animals rarely fossilize.

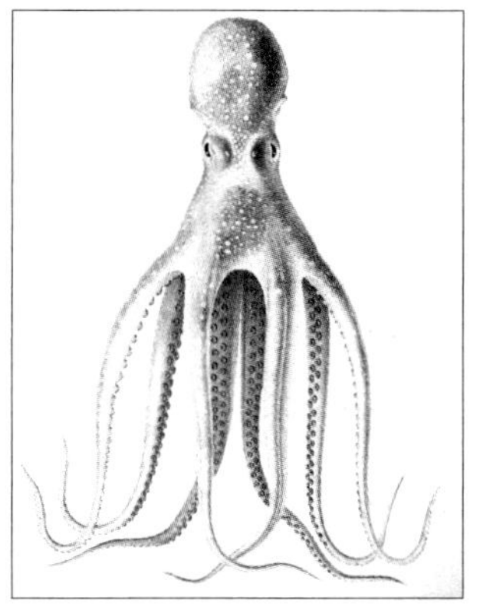

Octopus salulii Mediterranean Mollusk (1851).

COURTESY OF VINTAGE PRINTABLE

DISCOVERED	Hâdjoula (Lebanon)
SCIENTIFIC NAME	*Keuppia levante* Fuchs, Bracchi & Weis 2009
ETYMOLOGY	*Keuppia* in honor of paleontologist Helmut Keupp of Berlin; *levante* from the French word *levant*, meaning rising, as in the east, where the sun rises; also, the geographic and cultural region Levant, which includes Lebanon, Syria, Jordan, and Israel.
CLASSIFICATION	Animalia • Mollusca • Cephalapoda • Coleoidea • Palaeoctopodidae
PERIOD OF EXISTENCE	Late Cretaceous to Jurassic, 95 to 180 million years ago

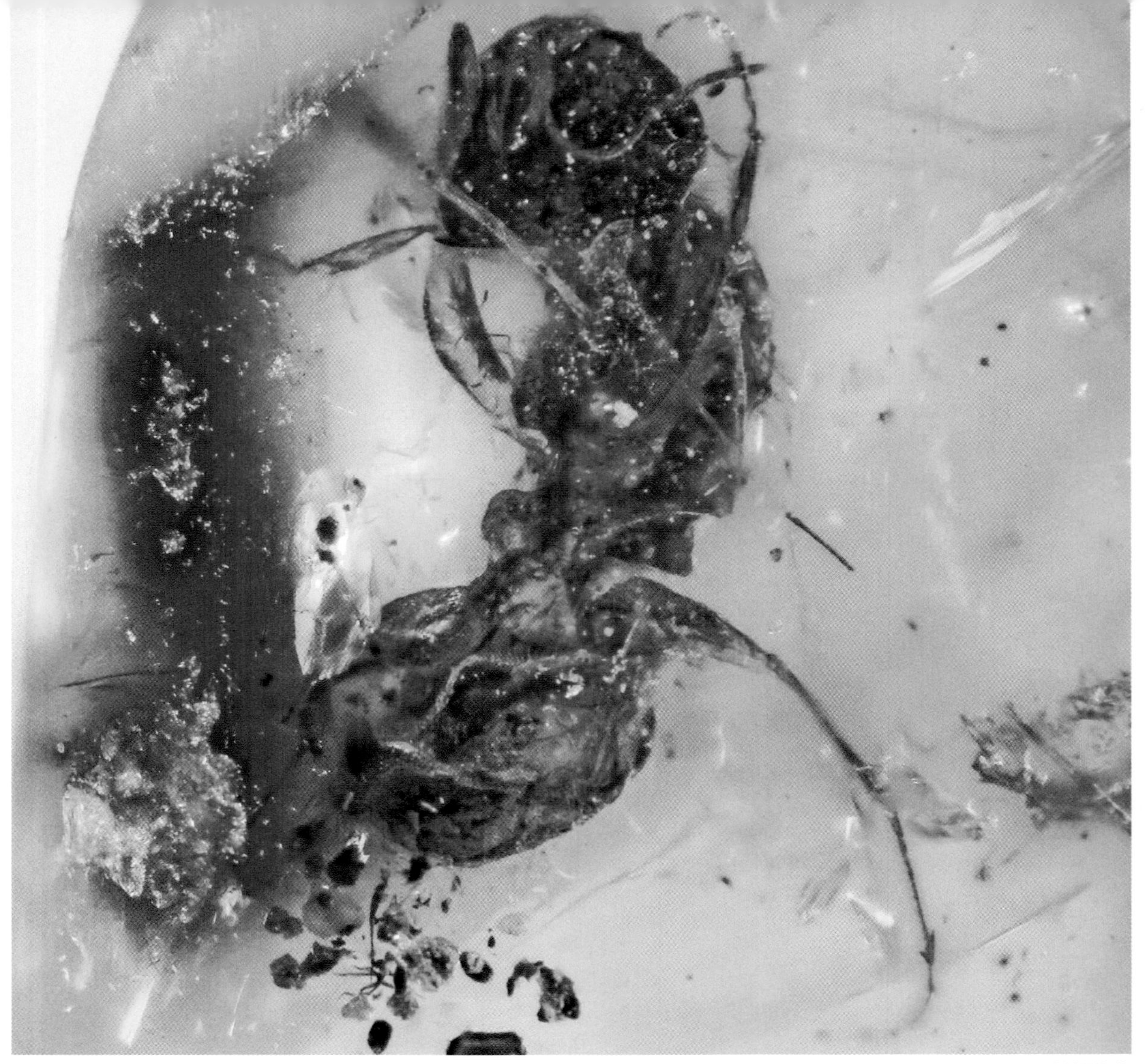

GEORGE POINAR JR.

Burma Bee

TO BEE OR NOT TO BEE

Forty million years older than the next-oldest known bee species is the 100-million-year-old *Melittosphex burmensis*. Measuring less than one-eighth of an inch, and with branched hairs holding grains of pollen, this ancient bee provides evidence that insect pollination contributed to the vast diversification and development of plant species during the Cretaceous Period. This significant fossil has attributes commonly associated with both bees and wasps, reinforcing the long-established view that bees are merely modified descendants of ancestral wasps.

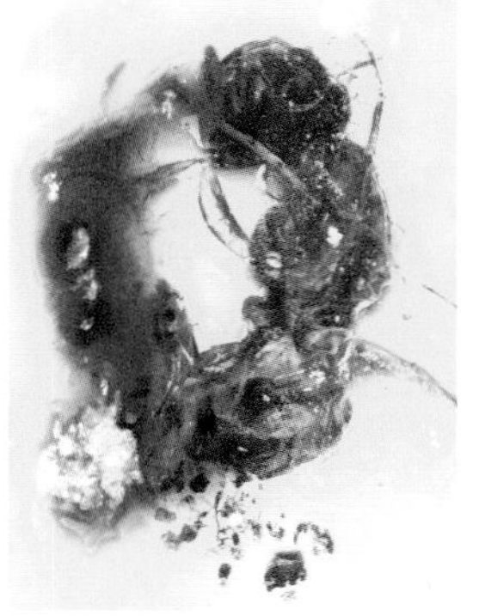

GEORGE POINAR JR.

Medieval swarm of bees

DISCOVERED	Amber mine in the Hukawng Valley, northern Myanmar (Burma)
SCIENTIFIC NAME	*Melittosphex burmensis* Poinar & Danforth 2006
ETYMOLOGY	*Melitto* from the Greek word for bee or honeybee, and *sphex*, also from the Greek, for wasp, indicating the fossil species link between bees and wasps; *burmensis* for the land where the fossil was found.
CLASSIFICATION	Animalia • Arthropoda • Insecta • Hymenoptera • Melittosphecidae
PERIOD OF EXISTENCE	Early Cretaceous, 100 million years ago

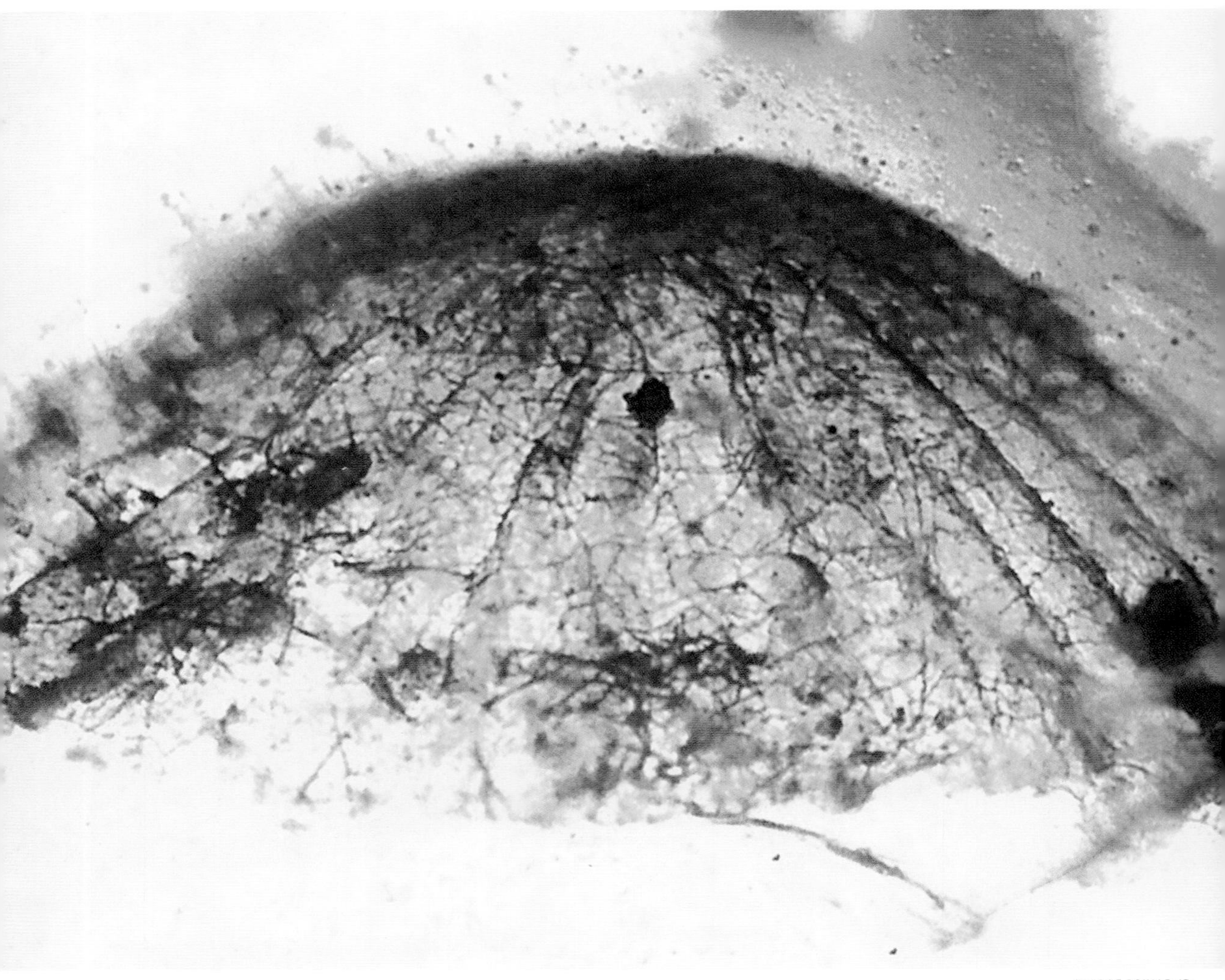

GEORGE POINAR JR.

Old-Old Mushroom

AGARIC AMBER ATTACK!

Only the fifth fossil mushroom ever to be discovered, *Palaeoagaracites antiquus* is believed to be the earliest fossil mushroom and only the second to be found this century. Dating to the Cretaceous Period and discovered in Burmese amber, *P. antiquus* is doubly significant because it sets a new minimum age for agarics—gilled mushrooms with caps and stems—and because this remarkable piece of amber provides 100-million-year-old evidence of one kind of fungus attacking another. In addition to *P. antiquus*, two other new fossilized fungus species were found in the same amber: *Mycetophagites atrebora* and *Entropezites patricii*. These represent two parasitic kinds of fungi, each attacking *P. antiquus*.

Agaricus strobiliformis, Mary Banning (1878).

COURTESY OF VINTAGE PRINTABLE

DISCOVERED	Amber mine in the Hukawng Valley, northern Myanmar (Burma).
SCIENTIFIC NAME	*Palaeoagaracites antiquus* Poinar & Buckley 2007
ETYMOLOGY	From the Greek words *paleos* for old and *agarikon* for fungus; *antiquus* from the Latin word meaning old or ancient.
CLASSIFICATION	Fungi • Basidiomycota • Agaricomycetes • Agaricales
PERIOD OF EXISTENCE	Early Cretaceous, 100 million years ago

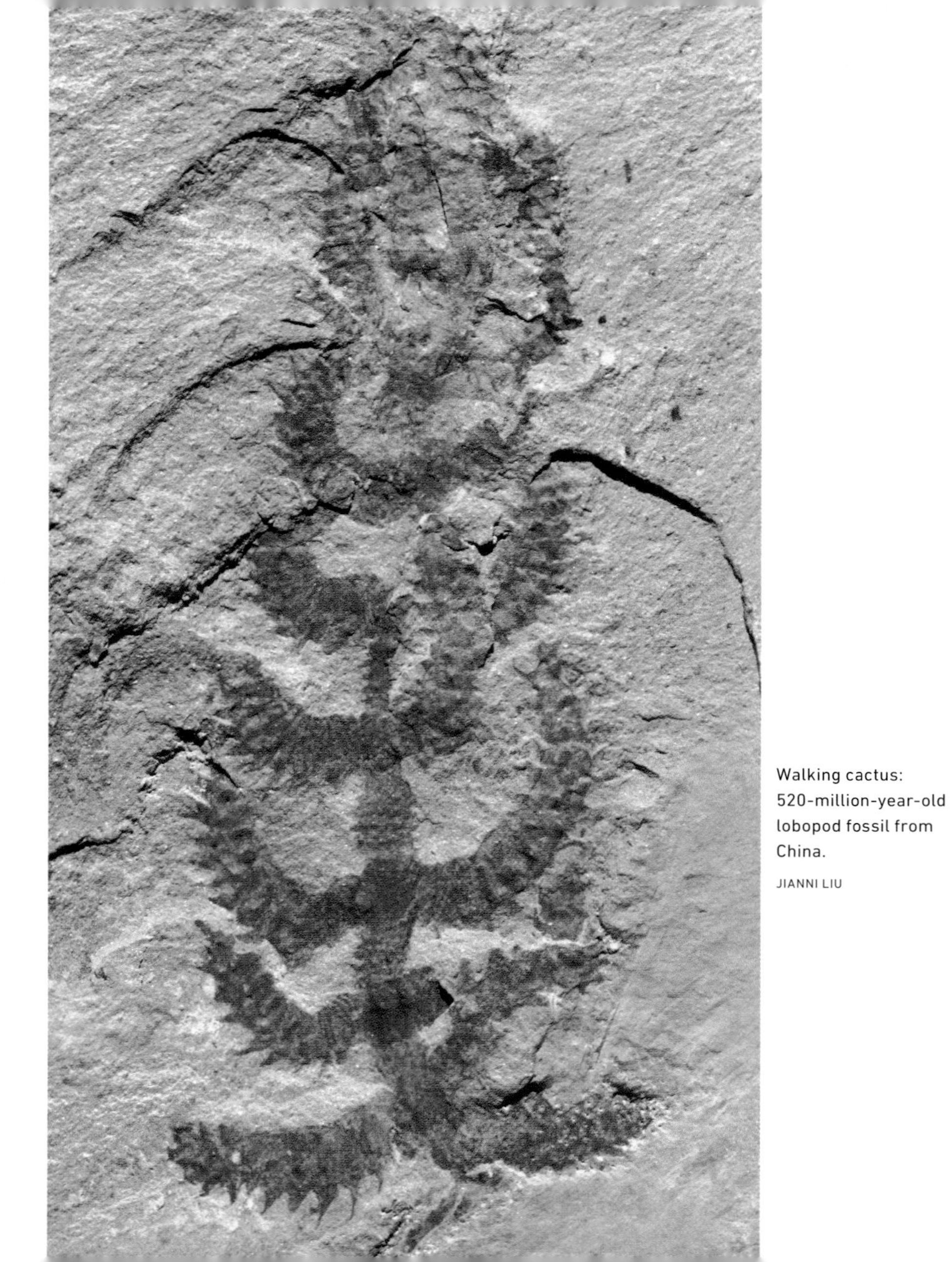

Walking cactus: 520-million-year-old lobopod fossil from China.

JIANNI LIU

Walking Cactus

ANCIENT ARTHROPOD?

Although this new species looks more like a cactus than an animal, *Diania cactiformis* belongs to an extinct group called the armored Lobopodia. Similar to today's living lobopodians—the Onychophora or velvet worms—the ancient armored lobopodians also had wormlike bodies and multiple pairs of legs. *D. cactiformis* is a significant discovery because its segmented legs add more evidence to the theory that arthropods (the largest group of living animals, including insects, spiders, and crustacea) evolved from lobopodian ancestors. Stated another way, it looks as if *D. cactiformis* may share a more recent common ancestor with arthropods than with other lobopodians—and in the scientific community, that is big news. *D. cactiformis* is about 2.4 inches long and was discovered in the famous Chengjiang deposit of southwest China.

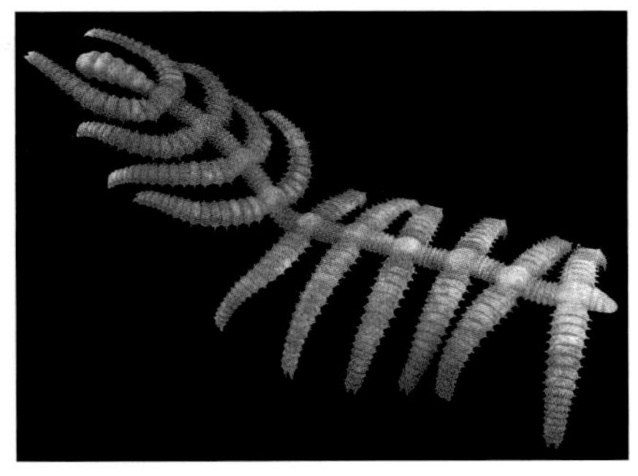

Reconstruction of *Diana cactiformis*.
JIANNI LIU

DISCOVERED	Yunnan, southwestern China
SCIENTIFIC NAME	*Diania cactiformis* Liu, Steiner, Dunlop, Keupp, Shu, Ou, Han, Zhang & Zhang 2011
ETYMOLOGY	*Diania* is named for Dian, a Chinese linguistic abbreviation of Yunnan, where the species was found; *cactiformis* refers to the animal's cactuslike form.
CLASSIFICATION	Animalia • Lobopodia • Xenusia
PERIOD OF EXISTENCE	Early Cambrian, about 520 million years ago

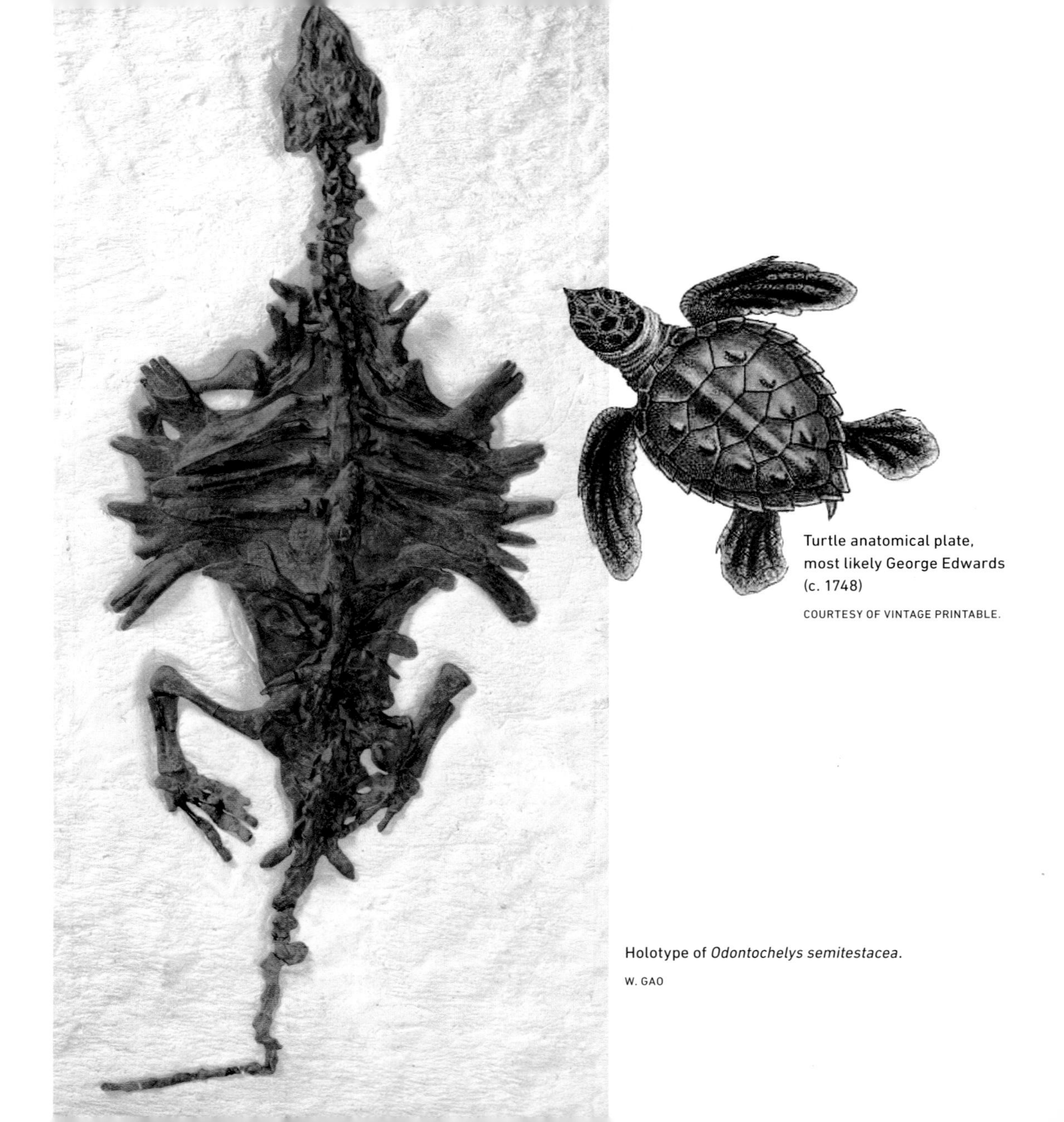

Turtle anatomical plate, most likely George Edwards (c. 1748)

COURTESY OF VINTAGE PRINTABLE.

Holotype of *Odontochelys semitestacea*.

W. GAO

Half-Shell Turtle

TURTLE ON THE HALF SHELL

Symbolizing longevity, patience, and wisdom, the turtle has held worldwide mythological status through time. In Chinese folklore, the creator goddess used the sea turtle to prop up the sky when the sea monster Gong Gong destroyed the pillar mountain that had supported the heavens. Given its sacred Chinese significance, it seems fitting that the oldest known turtle fossil, *Odontochelys semitestacea*, was discovered in China. The previously earliest known turtle, *Proganochelys*, had a shell that was fully formed on the top and bottom. *O. semitestacea* provides new evidence of the evolution of the turtle body form, since the belly portion of its shell had become developed but the top (dorsal) shell did not yet exist. The front legs of *O. semitestacea* are similar to those of currently living turtles that swim in shallow water, indicating a similar aquatic habitat for this early fossil.

Paratype of Odontochelys semitestacea.

W. GAO

DISCOVERED	Guizhou, China
SCIENTIFIC NAME	*Odontochelys semitestacea* Li, Wu, Rieppel, Wang & Zhao 2008
ETYMOLOGY	*Odont* from the Greek word for tooth, and *chelys* from the Greek *chelona* for tortoise; *semi* from the Latin prefix for half, and the Latin *testaceus*, meaning "covered with shell."
CLASSIFICATION	Animalia • Chordata • Reptilia • Testudinata • Odontochelyidae
PERIOD OF EXISTENCE	Late Triassic, 220 million years ago

COURTESY OF JIPB AND SPRINGER

Dila's Flower

CHINESE CORSAGE

Flowering plants and insects, coevolutionary partners, underwent explosive evolution and diversification during the Cretaceous Period. China's Yixian Formation is among the most productive on Earth and continues to yield some of the oldest fossils of flowers, including *Archaefructus* (1998), accepted by many botanists as the earliest known angiosperm. However, all parts of the "typical" modern flower did not evolve together. As a result, fascinating and unexpected combinations of features turn up among fossil plants and there is controversy over just where to draw the line for the origin of the first flowering plant. *Callianthus dilae* represents the world's oldest flower with a complete "modern" structure. The discovery of *C. dilae* and other Yixian fossil angiosperms has provided significant evidence that flowering plants by anyone's measure have existed on Earth for more than 125 million years.

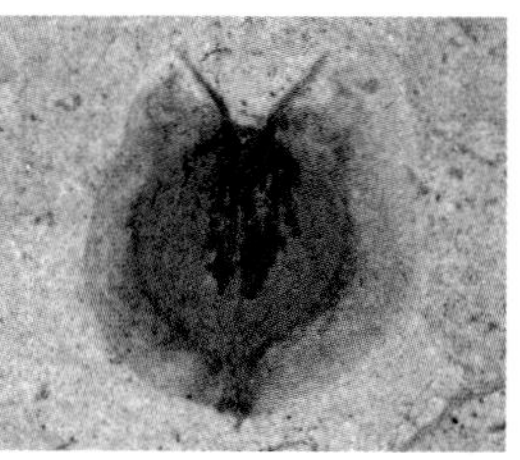

COURTESY OF JIPB AND SPRINGER

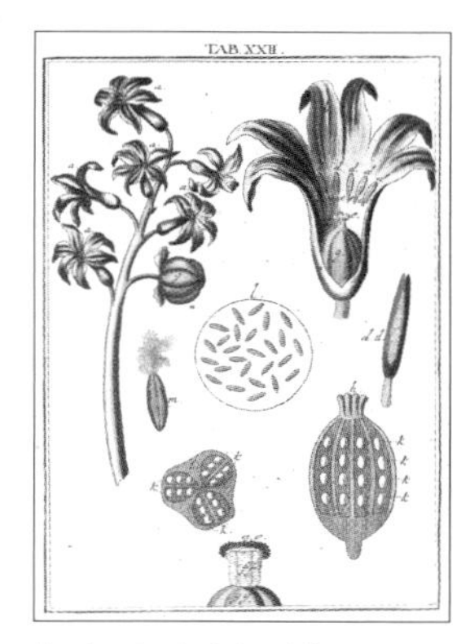

Anatomical plate of flower, Martin Frobenius (1764).

COURTESY OF VINTAGE PRINTABLE

DISCOVERED	Liaoning, China
SCIENTIFIC NAME	*Callianthus dilae* Wang & Zheng 2009
ETYMOLOGY	From the Greek words *kalos*, meaning beautiful, and *anthus*, meaning flower; *dilae* in honor of Mr. Dila Chen, who collected the fossil.
CLASSIFICATION	Plantae • Angiosperms
PERIOD OF EXISTENCE	Early Cretaceous, 125 million years ago

- **A** Tennessee Bottlebrush Crayfish
- **B** Martha's Pink Iquana
- **C** Pygmy Three-Toed Sloth
- **D** Cloudy Suckermouth Armored Catfish
- **E** Parecis Lizard
- **F** Isidoro's Chewing Lice
- **G** Giant Ribbed Clam
- **H** Matilda's Horned Viper
- **I** Strydom's Yam
- **J** Burrunan Dolphin

Chapter 6

HELLO, GOOD-BYE

THE MOST ENDANGERED NEW SPECIES

Over the course of the biosphere's 4-billion-year lifespan, our planet's geological record has chronicled five mass extinction events in which more than half of all species disappeared. As we noted in our chapter on new fossil species, the fifth and most recent extinction event occurred 65 million years ago, when a great meteor collided with Earth. Many environmental scientists believe that we are now entering the sixth great extinction event, and millions of species are in imminent danger of extinction. It has been suggested that the rate of extinction for this sixth event will be, on average, one thousand times greater than the recent average rate of extinction recorded by fossil evidence. However, comparisons are hindered by the fact that we know so little about the number and kind of species that currently exist. For the first time in human-recorded history, it is quite possible that species are disappearing more rapidly than they are being discovered and described.

And yet it's not all gloom and doom. Recent models indicate that we have more time to respond to the biodiversity crisis than previously thought. The speed at which Earth is losing its forests and jungles has slowed since the 1980s, when scientists sounded the alarm that connected habitat loss with species loss. There is also a growing recognition of the economic, ecological, and human importance of sustaining biodiversity. Plus,

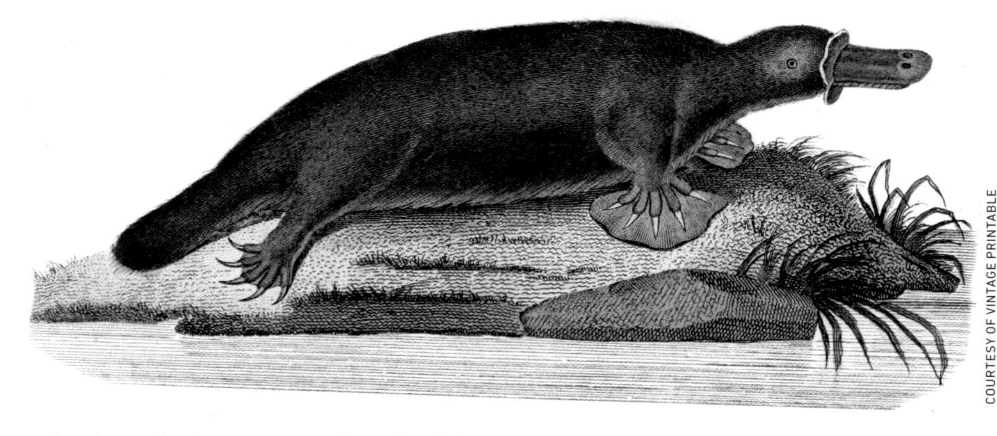

COURTESY OF VINTAGE PRINTABLE

technological advances—such as the ability to reach and photograph deep-sea trenches and the use of satellite imagery to identify new biodiversity hot spots—have made it possible to explore species more rapidly than ever. All of this suggests that if we capitalize on emerging technologies, then we can greatly accelerate the pace at which we discover species.

In order to detect, track, and respond to extinctions, it is imperative that we learn what species exist and where they are distributed in the biosphere. Most newly described species are known by only one or relatively few specimens collected from a handful of places. This often means that there is not enough data available to determine the conservation status of the species. Nonetheless, it is possible to infer the status of a poorly known species by correlating it with a unique habitat or ecosystem. In places like Madagascar or the Atlantic Coastal Forest of Brazil, 90 percent of the original forest cover has been destroyed, and we know that a high percentage of their species are found nowhere else on Earth. Any new species that is restricted to a habitat under stress is, by definition, threatened.

Regrettably, we had too many species candidates for this chapter, given the very large number of endangered or threatened species that have been newly discovered since 2000. Some of these new species are rapidly disappearing because they are being used as food sources by native populations, while others are losing their habitat due to land use pressures from industry, agriculture, or land development. Other new species are so clearly at risk upon discovery that scientists purposefully omit the location where the species was found in order to protect them. New species are often initially known from a tiny number of specimens, making it difficult to assess their status. When a newly discovered species is already critically endangered, there is no clearer demonstration of the biodiversity crisis. Hopefully, this chapter highlights the uncertain and potentially short future that many species face. Only by accelerating our exploration of species can we learn enough to make the right conservation decisions.

We should let the species in this chapter be a wake-up call to think about what is happening to the biosphere on our watch. There are still choices to be made, and we can either turn a blind eye to extinctions and hope for the best or choose to complete an inventory of species. By knowing where the wild things are, we become empowered to conserve and protect Earth's biodiversity.

COURTESY OF VINTAGE PRINTABLE

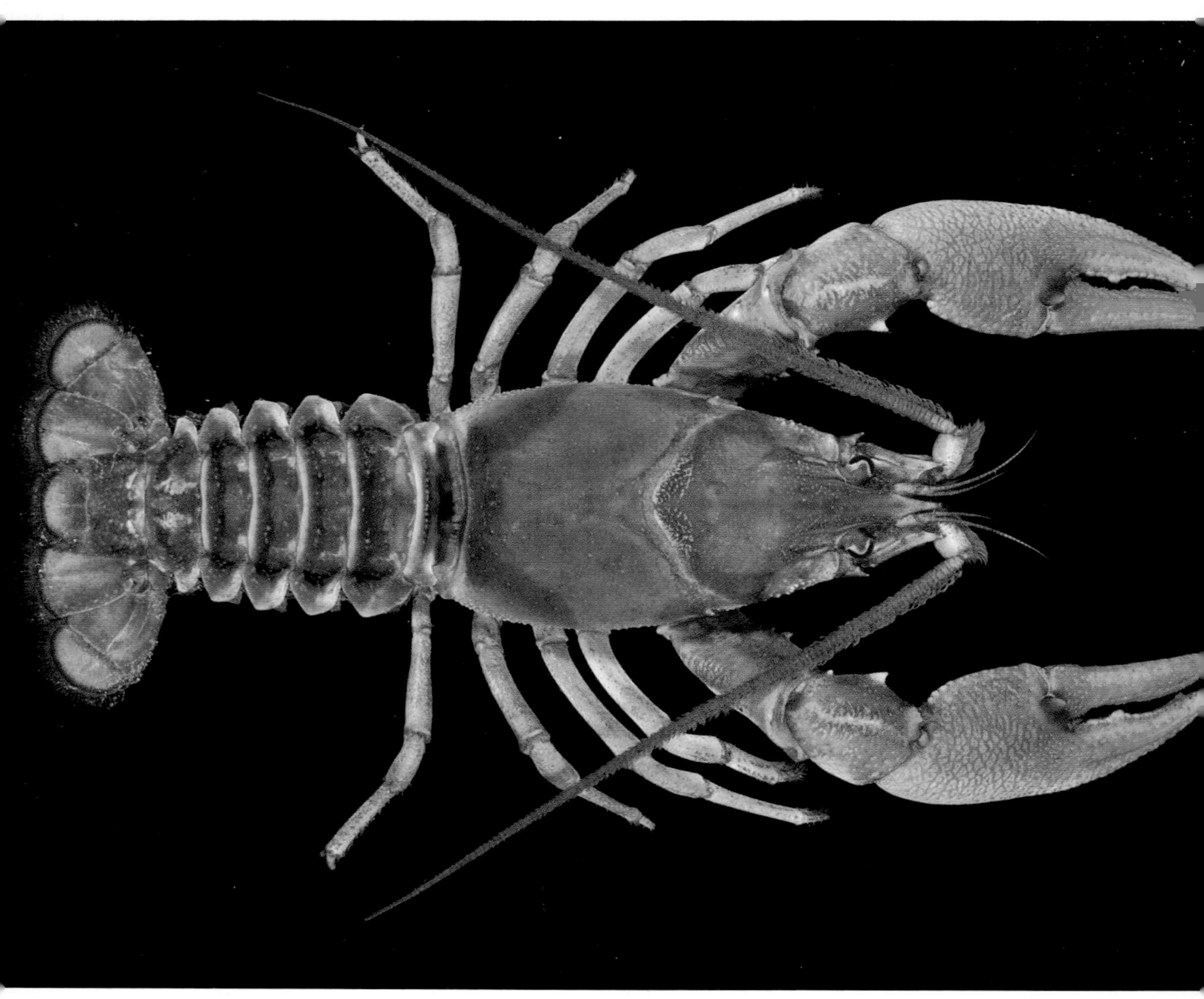

The Tennessee bottlebrush crayfish, *Barbicambarus simmonsi*, found in Factory Creek, Tennessee, in 2009.

C. E. WILLIAMS

Tennessee Bottlebrush Crayfish

CREEK-LOVIN' CRAWDAD

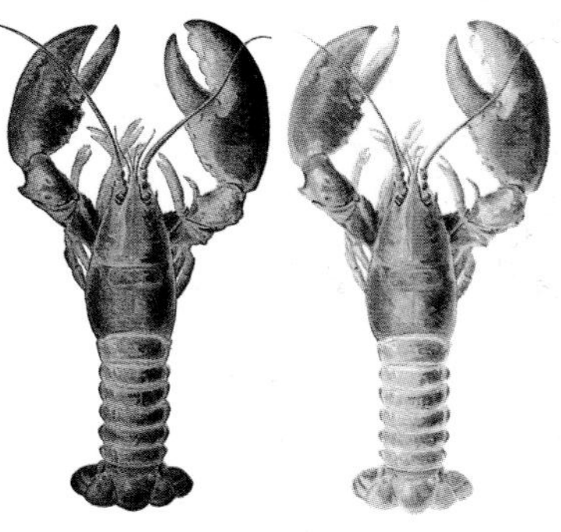

American lobsters, C. B. Beach (1904).

COURTESY OF VINTAGE PRINTABLE

Big as a lobster and incredibly rare, newly found crayfish *Barbicambarus simmonsi* is only the second species of its genus. The first species was a single, unusual crayfish with extraordinary "bearded" antennae discovered near Mammoth Cave, Kentucky, in 1884 and originally named *Cambarus cornutus* (later changed to *B. cornutus*). Although hundreds of new crayfish species have been discovered around the world over the past 128 years, none of them have come close to *B. cornutus* and it remained unique until 2009, when the gigantic *B. simmonsi* was found in Tennessee's Shoal Creek. Crayfish of the southeastern United States have been intensively studied for decades, making *B. simmonsi*'s discovery totally unexpected. Despite its size, and extensive exploration, only five specimens of *B. simmonsi* have been collected, leading scientists to believe that this new species is most likely vulnerable, given its tiny population and microhabitat.

DISCOVERED	Tennessee, USA
SCIENTIFIC NAME	*Barbicambarus simmonsi* Taylor & Schuster 2010
ETYMOLOGY	*Barbicambarus* is a genus of crayfish named in 1969; *simmonsi* is named in honor Jeffrey W. Simmons, an aquatic biologist with the Tennessee Valley Authority, who discovered the species.
CLASSIFICATION	Animalia • Arthropoda • Crustacea • Malacostraca • Decapoda • Cambaridae

GABRIELE GENTILE

Martha's Pink Iguana

DARWIN DO-OVER?

GABRIELE GENTILE

During his 1835 voyage to the Galápagos Islands on the survey ship *Beagle*, Charles Darwin became inspired by the animals and plants that were unique to the archipelago. Upon his return to England, he developed his theory of evolution and wrote his famous book *On the Origin of Species*. At the time, however, he didn't appear to be terribly impressed with the variously yellow-orange, brown, and red Galápagos land iguanas, which he described as "ugly animals" having a "singularly stupid appearance." If he had only seen rosy *Conolophus marthae* he may have been tickled pink. Discovered in 2006, *C. marthae* is now considered to be critically endangered due to its tiny population of about forty-two animals; a highly restricted habitat that is limited to less than ten square miles on a single, small island; and risk factors associated with disease and environmental dangers related to volcanic activity and predators.

DISCOVERED	Galápagos National Park, Ecuador
SCIENTIFIC NAME	*Conolophus marthae* Gentile & Snell 2009
ETYMOLOGY	*Conolophus* is a genus of iguanas found only on the Galápagos Islands and was named in 1831; *marthae* is named in memory of Gabriele Gentile's daughter, Martha Rebecca Gentile.
CLASSIFICATION	Animalia • Chordata • Reptilia • Squamata • Iguanidae

BRYSON VOIRIN

Pygmy Three-Toed Sloth

A SHRINKING BRADY BUNCH

© CRAIG TURNER / ZSL

About ten thousand years ago, ocean levels began to rise in the Americas, causing an island archipelago to form near present-day Panama as water separated high areas from the mainland. The oldest and most remote island in this group is the Isla Escudo de Veraguas—the only environment where new species *Bradypus pygmaeus* can be found. In addition to being restricted to such a unique and limited habitat, this new sloth has a diminishing population of about one hundred individuals and is considered to be critically endangered due to land development and the impact of tourism. Fortunately, *B. pygmaeus* avoids detection because it stays in trees, where it hardly moves, and its fur turns green with algae, providing camouflage from predators. As its name implies, *B. pygmaeus* is small. Compared with Panama's mainland sloth *B. variegatus*, this pygmy sloth is 40 percent smaller in mass and 15 percent shorter in body length.

DISCOVERED	Panama
SCIENTIFIC NAME	*Bradypus pygmaeus* Anderson & Handley 2001
ETYMOLOGY	*Bradypus* is a genus of three-toed sloths named by Linnaeus in 1758; *pygmaeus* is from the Latin meaning pygmy or dwarf.
CLASSIFICATION	Animalia • Chordata • Mammalia • Pilosa • Bradypodidae

NADIA MILANI / CARLOS DONASCIMIENTO

Cloudy Suckermouth Armored Catfish

VENEZUELAN VENI VIDI VICI

NADIA MILANI / CARLOS DONASCIMIENTO

Completed in 2009 and funded by the U.S. National Science Foundation, the *All Catfish Species Inventory* was one of the most expansive Planetary Biodiversity Inventory projects demonstrating that large international teams of scientists and museums could accelerate the rate of new species discovery. The effort to inventory Earth's species is especially important considering that many undescribed species may be on the brink of extinction, such as *Cordylancistrus nephelion*. Found only in the Tuy River of north-central Venezuela, *C. nephelion* is critically endangered due to the river's "extreme alteration" as a result of dam construction and heavy pollution from industry and human-produced wastewater. *C. nephelion* is just one of 430 of the project's new catfish species and is distinctive in its genus for its unique coloration of white irregular spots on a greenish, black-brown body. Suckermouth fish have mouths on their bellyside that have adapted for "grazing" on food and algae growing beneath them.

DISCOVERED	Tuy River, north-central Venezuela
SCIENTIFIC NAME	*Cordylancistrus nephelion* Provenzano & Milani 2006
ETYMOLOGY	*Cordylancistrus* is a genus of catfish described in 1980; *nephelion* is from the Greek *nephele*, meaning cloud, in reference to the fish's irregular, cloudlike spots.
CLASSIFICATION	Animalia • Chordata • Actinopterygii • Siluriformes • Loricariidae

GUARINO RINALDI COLLI

Parecis Lizard

BECOMING A HAS-BEEN

Discovered near the city of Vilhena, Brazil, and described in 2003, new species *Cnemidophorus parecis* is considered critically endangered due to loss of habitat where only 10 percent of Vilhena's native vegetation remains, making it one of Amazonia's most deforested areas. *C. parecis* lives on open, grassy ground and under shrubs in a very restricted range of Brazil's savanna, the Cerrado. This new species produces a small clutch (the number of eggs laid at one time) of one or two eggs. According to a recent study, clutch size is "a good predictor of extinction" for lizards that do not live on islands and whose habitat is threatened. Continued encroachment on the lizards' habitat has followed recent state legislation that allows for more rural land to be converted to large agribusiness properties and a consequent boost in land given over to soybean production.

GUARINO RINALDI COLLI

DISCOVERED Cerrado, Brazil

SCIENTIFIC NAME *Cnemidophorus parecis* Colli, Costa, Garda, Kopp, Mesquita, Péres Jr., Valdujo, Vieira & Wiederhecker 2003

ETYMOLOGY *Cnemidophorus* is a genus of lizards named in 1830; *parecis* is in reference to the indigenous Paresi people and to the Brazil highlands, Chapada dos Parecis, where the species was found.

CLASSIFICATION Animalia • Chordata • Reptilia • Squamata • Teiidae

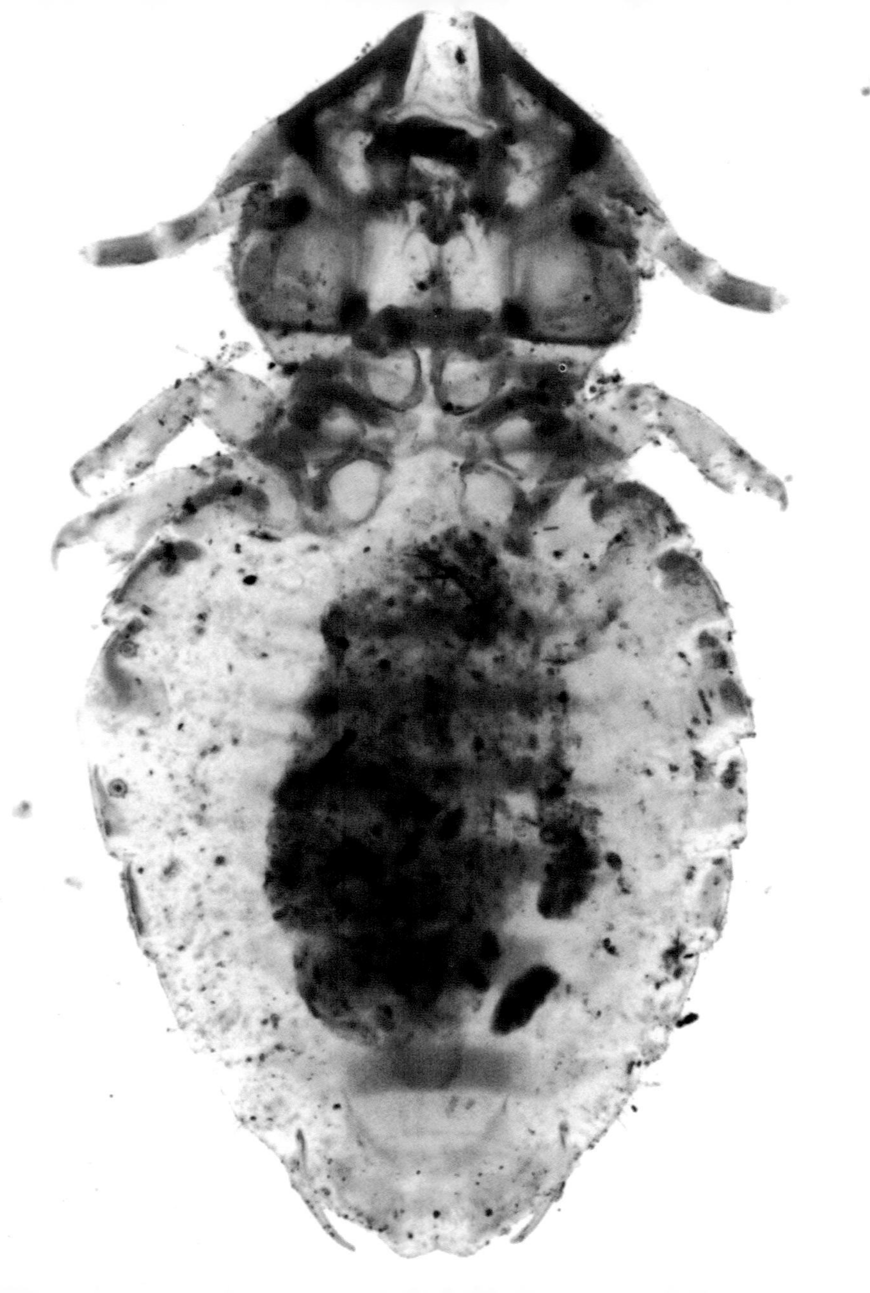

A chewing louse similar to new species *Felicola isidoroi*.

JESUS M. PEREZ

Isidoro's Chewing Louse

A LOUSY DEAL FOR BOTH

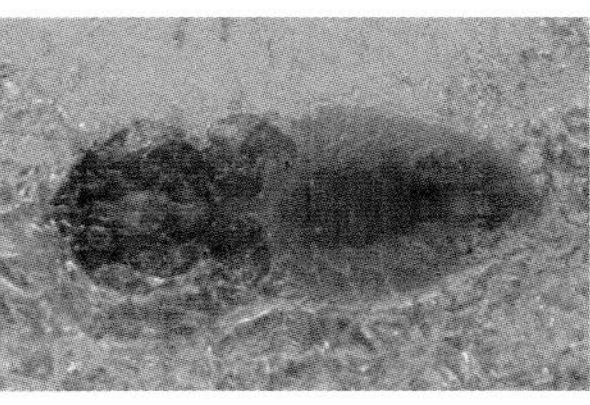

JESUS M. PEREZ

Not to be nitpicky, but lice are some of the pickiest animals in the world—to the point that some lice will only feed on just one species, a phenomenon known as host specificity. Unfortunately for lice, if their host is endangered, then they suffer "habitat" loss and are endangered, too. The Iberian lynx, *Lynx pardinus*, is among the most endangered wildcats in the world, and thus, the fate of a host-specific louse species on the Iberian lynx does not bode well. Put simply, if the Iberian lynx becomes extinct, so does *Felicola isidoroi*. Parasites play a role in regulating host populations, showing us that every species, even lousy ones, can teach us about evolution and ecology. Like an evolutionary version of the film *It's a Wonderful Life*, this example of *L. pardinus* and its partner louse reminds us that every species touches many others, and that one extinction often begets another. If we can save the Iberian lynx, then we can also ensure the survival of this endangered louse.

Iberian lynx, critically endangered host of Isidoro's chewing louse.

DISCOVERED	Spain
SCIENTIFIC NAME	*Felicola isidoroi* Perez & Palma 2001
ETYMOLOGY	*Felicola* is a genus of lice named in 1929; *isidoroi* is named in memory of Dr. Isidoro Ruiz-Martínez.
CLASSIFICATION	Animalia • Arthropoda • Insecta • Mallophaga • Trichodectidae

Giant Ribbed Clam

STARCLAM CALAMITY

Legends have spoken of giant clams as "man-eating" killers that would grasp a sailor's arm between their mighty shells and drown him in the sea. More than 125,000 years ago, *Tridacna costata* was one of the most prolific and accessible mollusks found in the Red Sea; it accounts for over 80 percent of the fossil shells found in the area. However, this giant clam has fallen on hard times and predator has become prey. Only thirteen live specimens of *T. costata* were collected in describing this new species, due to its current scarcity as a result of humans gathering it for food. Discovered in 2006, this rare—and perhaps critically endangered—giant can weigh up to four and a half pounds, and its distinctive zigzag shell, with five to seven vertical folds, can be two feet in length.

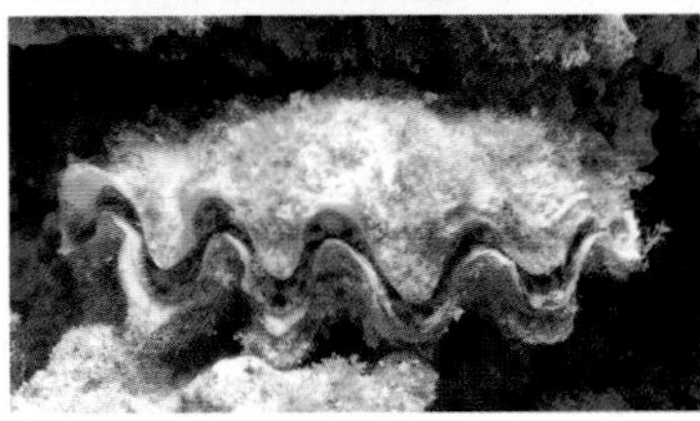

DISCOVERED	northeastern Gulf of Aqaba, Jordanian Red Sea
SCIENTIFIC NAME	*Tridacna costata* Roa-Quiaolt, Kochzius, Jantzen, Zibdah & Richter 2008
ETYMOLOGY	*Tridacna* is a genus of giant clams named in 1797; *costata* from the Latin *costa*, meaning rib, for the clam's riblike folds.
CLASSIFICATION	Animalia • Mollusca • Bivalvia • Veneroida • Tridacnidae

MICHELE MENEGON

Matilda's Horned Viper

TANZANIAN DEVIL

MICHELE MENEGON

It's been theorized that over the millennia, humans have developed highly tuned snake-detection skills as an adaptation for avoiding poisonous vipers. Although *Atheris matildae* is brightly colored and over two feet long, the possibility of anyone seeing this critically endangered—and dangerous—snake is remote. Only a tiny number of *A. matildae* have been found in a hundred-square-kilometer area (38.6 square miles) in the isolated forests of Tanzania's Southern Highlands. Not only is the population of *A. matildae* small, but the snake's habitat is becoming smaller as the forests become interspersed with grasslands and increasingly fragmented from the broader, more sustainable forest environment. *A. matildae* was first discovered in 2009 and ten other snakes were collected in 2011 for a conservation breeding program as insurance against exploitation and loss of habitat. Other species that live only in these forest fragments are critically endangered as well, including the recently discovered kipungi monkey (2005).

DISCOVERED	Southern Highlands, Tanzania
SCIENTIFIC NAME	*Atheris matildae* Menegon, Davenport & Howell 2011
ETYMOLOGY	*Atheris* is a genus of venomous vipers named in 1862; *matildae* is named in honor of Matilda Davenport, daughter of Tim Davenport, one of the discoverers.
CLASSIFICATION	Animalia • Chordata • Reptilia • Squamata • Viperidae

JOHN BURROWS

Strydom's Yam

MEDICINE YAM

In many parts of the world, yams are a vital food source and at least one new species of edible yam, *Dioscorea orangeana* (2009), has become threatened due to overharvesting. Some yams, including new species *D. strydomiana*, are being exploited for their medicinal value. Since 2002, several conservation groups have raised concerns about threats to *D. strydomiana*, and considerable efforts have been made to monitor plant collection and to bank its seeds. Only two hundred plants were found in 2008 and, of those, almost 89 percent showed evidence of some destruction. Not only is *D. strydomiana* critically endangered through overcollection for medicine, but this yam is slow growing, has a restricted habitat, and is susceptible to a host of environmental risks that include agriculture, mining, and being damaged by porcupines.

JOHN BURROWS

DISCOVERED	South Africa
SCIENTIFIC NAME	*Dioscorea strydomiana* Wilkin 2010
ETYMOLOGY	*Dioscorea* is a genus of flowering plants named by Linnaeus in 1753; *strydomiana* is named in honor of Gerhard Strydom for his significant role in the species' discovery.
CLASSIFICATION	Plantae • Angiosperms • Monocots • Dioscoreales • Dioscoreaceae

ADRIAN HOWARD

Burrunan Dolphin

THE NOSE HAS IT

KATE CHARLTON-ROBB

Given the rarity of the discovery of new mammal species—especially large-sized mammals—the documentation of a new dolphin species is truly significant. *Tursiops australis* is an especially exciting discovery because these Australian dolphins from the Victorian coast were long believed to be a common bottlenose dolphin, *T. aduncus* or *T. truncatus*. To the amazement of the scientists conducting a genetics study, the dolphins turned out to be neither. There are currently only two very small populations of *T. australis*, with a total count of about 170 dolphins. These dolphins are found nowhere else in the world and are considered to be endangered, with fears brewing in the scientific community that this species may not persist. In general, worldwide dolphin populations are declining because of fishing-net kills and through habitat destruction as a direct result of tourism, coastline development, and industrial wastewater pollution.

DISCOVERED	southeastern coastal waters of Australia
SCIENTIFIC NAME	*Tursiops australis* Charlton-Robb, Gershwin, Thompson, Austin, Owen & McKechnie 2011
ETYMOLOGY	*Tursiops* is a genus of dolphins named in 1855; *australis* is both the Latin for southern and in reference to Australia, where the dolphin lives.
CLASSIFICATION	Animalia • Chordata • Mammalia • Cetacea • Delphinidae

A Leprosy Bacterium

B Doris Swanson's Poison Dart Frog

C Corredor's Assasin Bug

D Zombie Ant Fungus

E Lilian's Widow Spider

F Ashe's Cobra

G Attenborough's Pitcher

H Andre Menez's Cone Snail

I Poisonous Predaceous Polyclad

J King's Deadly Jelly

Chapter 7

LETHAL WEAPONS: VENOMS, TOXINS, AND DISEASE

THE DEADLIEST NEW SPECIES

It's 1967: the classically campy TV series *Lost in Space* is in its third season and broadcasting its eleventh episode, "Deadliest of the Species." Viewers were warned: Danger, Will Robinson! Danger! Although this popular catchphrase was used only once on the show, it's still good advice if you're conducting a scientific expedition on planet Earth and discover new species such as the deadly examples in this chapter.

Most people associate danger with nature's usual suspects—swooping birds of prey, big-toothed carnivores, or "man-eating" sharks. There are many times when species advertise the danger they present. Rattlesnakes give an audible warning before striking. Many bees and wasps have highly visible stripes so that potential predators will learn to steer clear once stung. But danger can come from a wide diversity of species. In this chapter, we've chosen new species that are dangerous for unusual reasons that will

Death cap mushroom, *Amanita phalloides*, plate 35, Joseph Sturm (1817).

COURTESY HARVARD UNIVERSITY BOTANY LIBRARIES

stretch your conception of what constitutes a potentially lethal species. Some may cause or carry disease while others may use powerful toxins to defend themselves or overpower their prey. Appearances can be deceiving as well—many lethal species are benign-looking or even pretty, such as a colorful poison dart frog, an unassuming pitcher plant, or a beautifully patterned cone snail. Most of the species in this chapter are small in size yet produce venoms or poisons that are deadly to other species.

It is worth noting the distinction between species that are poisonous and those that are venomous. Venomous animals produce toxins that are injected into a victim and used primarily for offense. Poisons play for the defense. Poisonous animals deliver their punch through touch or ingestion and often acquire toxicity from a food source. For example, the toxins found in the skin of poison dart frogs in the wild are developed from the frogs' primary diet of insects. The same species of poison frog raised in captivity produces no toxins since their food source no longer provides it. The same may be said of poisonous insects that acquire toxic chemicals from host plants. Poisonous plants, on the other hand, grow their own.

Danger is relative. Something benign to one kind of plant or animal may present the worst imaginable fate to another. Many disease-causing fungi, for example, are very specific in the organisms they attack and may be deadly to just one species. Some snake species can eat poison dart frogs with no effect, while the same dosage of the poison could kill twenty humans. Parasites—or disease vectors—may present no consequence to the animals that carry them yet infect other animals, causing grave illness or death. In most cases, humans are collateral damage in this world of biochemical warfare. Wasps and bees evolved potent stings to defend their nests, not to inflict pain or cause deadly anaphylactic shock in people. There are some mushrooms that can be eaten by one person but make another deathly ill. Poison ivy has noxious chemicals to protect itself from insect predators that also cause a severe allergic reaction in some humans.

Caution is a better watchword than fear when going into nature. For every species that presents a serious threat of death to humans, there are millions that do not. And with a little common sense, most potentially dangerous encounters can be avoided or minimized. While poisoning, stinging, biting, and infecting other organisms can seem like a destructive force of nature, it is also just the opposite: yet another example of natural selection—the positive force behind the evolution of new species.

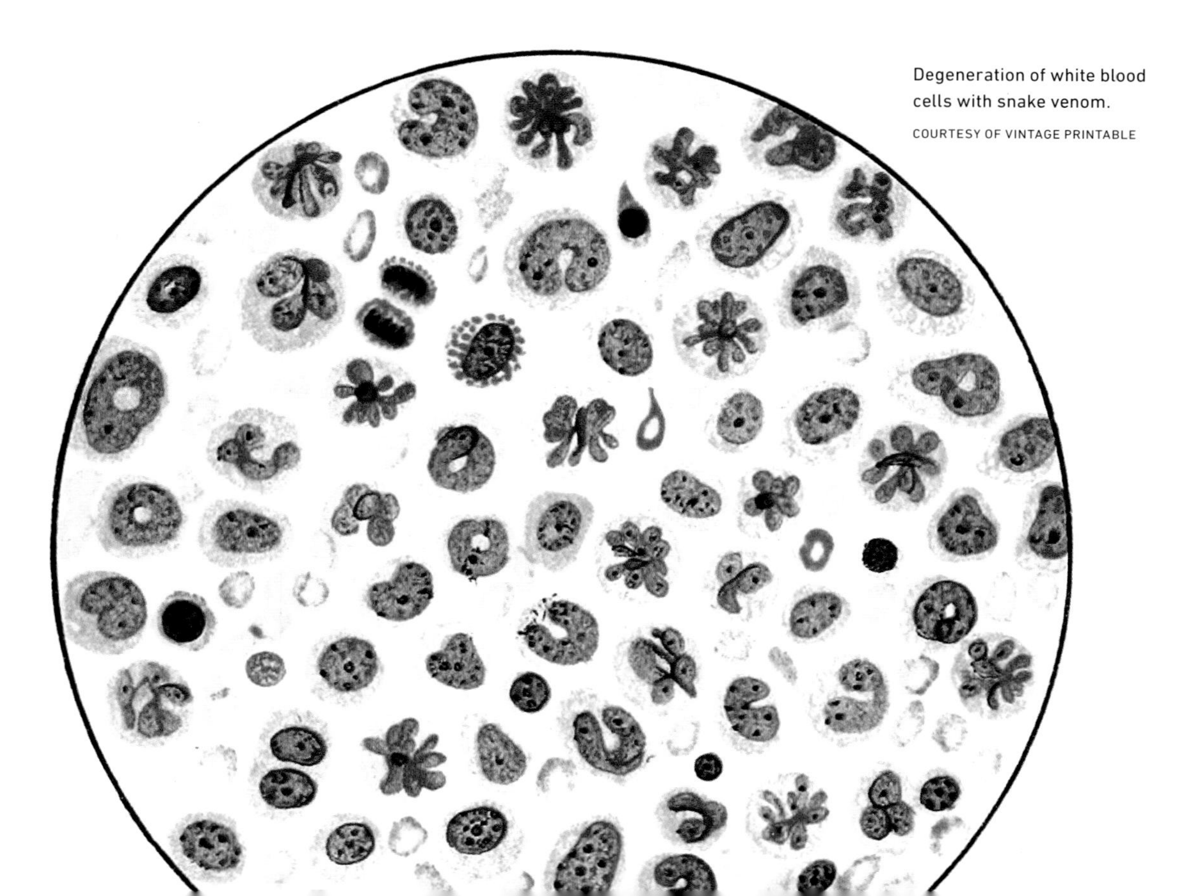

Degeneration of white blood cells with snake venom.

COURTESY OF VINTAGE PRINTABLE

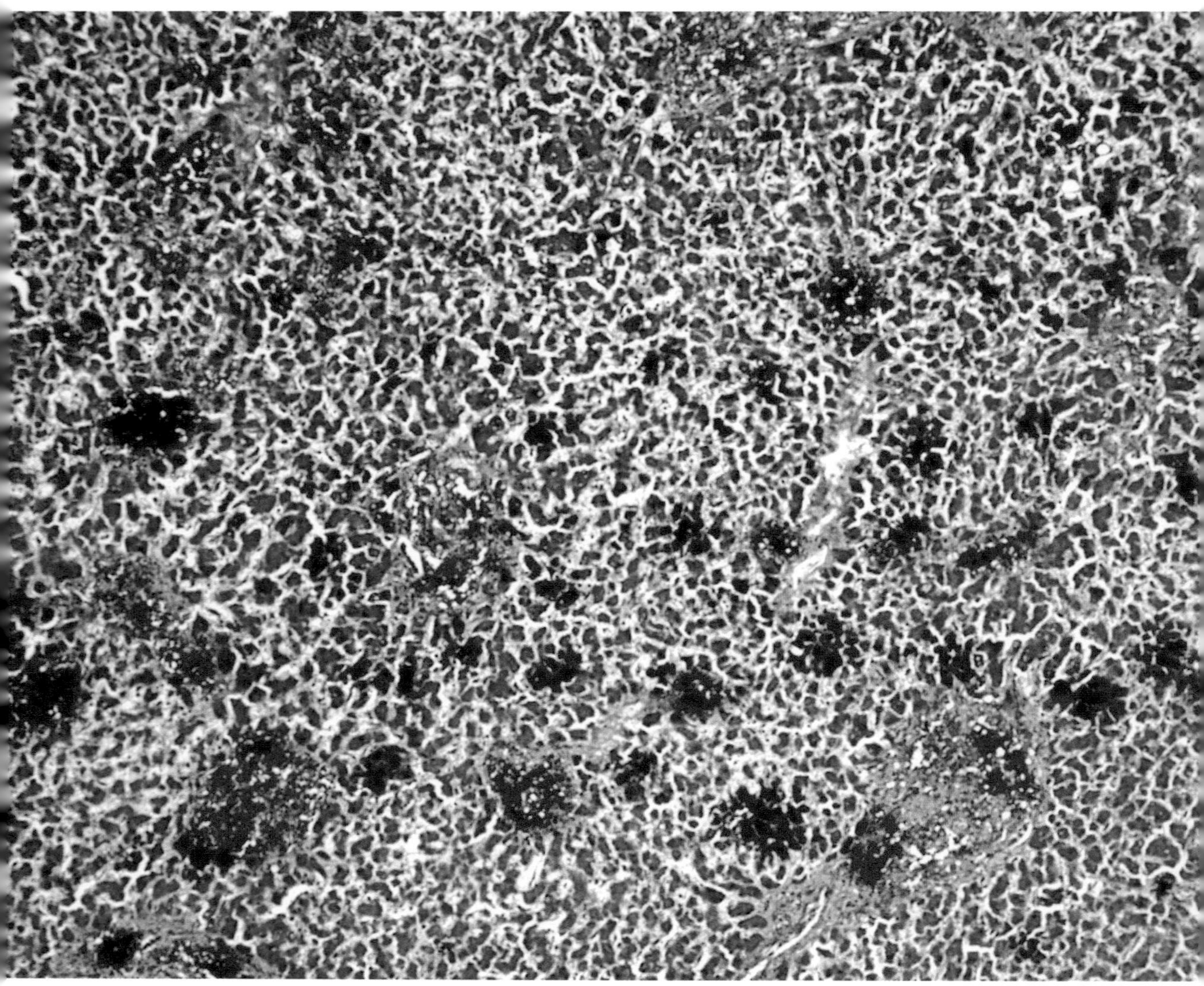

Leprosy Bacterium

A LEPER WITH NEW SPOTS

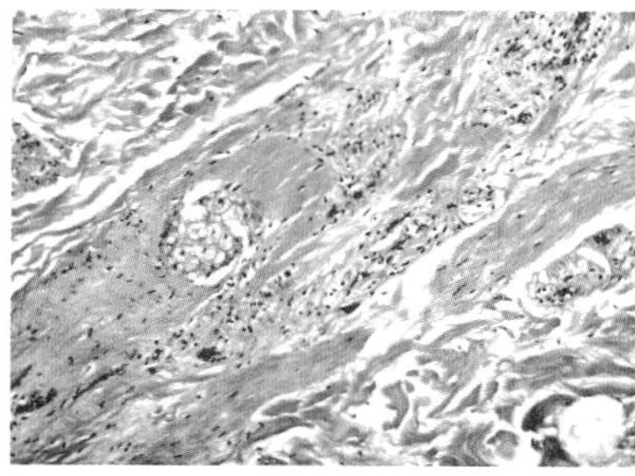

© AMERICAN SOCIETY FOR CLINICAL PATHOLOGY

Leprosy (Hansen's disease) is a nerve-damaging and disfiguring disease that has afflicted humans for thousands of years. Almost never fatal, leprosy continues to permanently disable hundreds of thousands of people each year. Until recently, most cases were known or assumed to be caused by *Mycobacterium leprae*, a bacterium first isolated by G. A. Hansen in 1873. In 2007, a new species of bacterium was discovered that causes the most severe form of the disease, diffuse lepromatous leprosy. *M. lepromatosis* was found in the systems of two Arizona residents following their deaths and was also subsequently encountered in two fatal cases in Singapore. Scientists are working to establish whether the new species is responsible for the more severe form of leprosy in other patients. The discovery of this new and deadly bacterium may help explain geographic differences among forms of the disease as well as improve its diagnosis and treatment.

DISCOVERED	USA, from two case studies in a Phoenix, Arizona, hospital and analyzed in Texas
SCIENTIFIC NAME	*Mycobacterium lepromatosis* Han, Seo, Sizer, Schoberle, May, Spencer, Li & Nair 2008
ETYMOLOGY	*Mycobacterium* is a genus of bacteria named in 1896; *lepromatosis* is from the Greek words *lepros*, meaning scaly; *lepein*, meaning to peel; and *osis*, meaning "affected with."
CLASSIFICATION	Bacteria • Actinobacteria • Actinomycetales • Mycobacteriaceae

COURTESY OF TARAN GRANT

Doris Swanson's Poison Dart Frog

KERMIT'S ALTER EGO

MAURICIO RIVERA-CORREA

The beautiful bright colors of poison dart frogs are a warning of potentially paralyzing and sometimes deadly toxins to potential predators. One class of toxins, batrachotoxins, is especially dangerous and includes some of the most lethal and persistent poisons in the world. A single poison dart frog can produce enough batrachotoxin to kill twenty thousand mice, and when applied to a dart used in hunting, the toxin can retain its potency for three years. Only a few species of poison dart frogs are deadly, but all are dangerous. The new species *Dendrobates dorisswansoni* most likely secretes pumiliotoxins that, although significantly weaker than batrachotoxins, are nonetheless an effective defense, disrupting signals in the nervous and skeletal muscle systems. Although pumiliotoxins are poisonous in high concentrations, scientists are finding medical applications for them, including use as a stimulant following a heart attack.

DISCOVERED	Bosques de Florencia National Natural Park, Colombia
SCIENTIFIC NAME	*Dendrobates dorisswansoni* Rueda-Almonacid, Rada, Sánchez-Pacheco, Velásquez-Álvarez & Quevedo 2006
ETYMOLOGY	*Dendrobates* is a genus of frogs named in 1830; *dorisswansoni* in honor of Doris Swanson of Spokane, Washington, for her dedication to conservation and biodiversity.
CLASSIFICATION	Animalia • Chordata • Amphibia • Anura • Dendrobatidae

CLEBER GALVÃO & VICTOR M. ANGULO

Corredor's Assassin Bug

COLOMBIAN BUG CARTEL

Assassin bug illustration (1877).

COURTESY OF VINTAGE PRINTABLE

Belminus corredori, a new species from Colombia, joins a growing list of more than 130 South American species of triatomine bugs responsible for the spread of Chagas' disease, which kills about twenty-thousand people per year. Also known as assassin bugs, these insects don't directly cause Chagas' disease but transmit the parasite species that causes the disease and is found in the feces of the assassin bug. People living in poverty and poorly sealed huts are susceptible to being bitten by these insects as they sleep. A slumbering victim instinctively scratches the itchy site of the bite, smearing bug feces and parasites into the wound, introducing infection. Many disease vectors are arthropods—such as insects—although other animals can be vectors or intermediate hosts. The most commonly known disease vector is probably the bloodsucking mosquito that carries the parasites that cause malaria.

DISCOVERED	Santander, Colombia
SCIENTIFIC NAME	*Belminus corredori* Galvão & Angulo 2006
ETYMOLOGY	*Belminus* is a genus of true bugs named in 1859; *corredori* is in honor of Dr. Augusto Corredor Arjona, for his contributions to the fields of parasitology and public health and his development of tropical disease control programs in Colombia.
CLASSIFICATION	Animalia • Arthropoda • Insecta • Hemiptera • Reduviidae

DAVID P. HUGHES

Zombie Ant Fungus

PLIGHT OF THE LIVING DEAD

Ophiocordyceps camponoti-rufipedis is one of four newly discovered fungus species that reproduce by infecting ants and turning them into zombies. Details of the body snatching are not yet known, but, once infected, the ants behave abnormally and leave their nest. They clamp their jaws onto leaves, sprout fruiting fungal bodies from their necks, and spread zombie-making spores to other ants. Each of the four new zombie fungus species from Brazil's Zona da Mata Atlantic forest system attacks a different species of carpenter ant. The fungus experts who discovered these species are concerned that a drying climate in the high-elevation home of the fungi may threaten their survival—which may come as good news for ants. The good news for the scientists who study fungi? It's likely that these new species are just a small fraction of zombie ant fungus diversity waiting to be discovered.

Carpenter ant *Camponotus rufipes* before fungal infection.

Carpenter ant *Camponotus rufipes* after fungal infection.

DAVID P. HUGHES

DISCOVERED	in 2010 in the Atlantic rain forests of Minas Gerais, Brazil
SCIENTIFIC NAME	*Ophiocordyceps camponoti-rufipedis* Evans & Hughes 2011
ETYMOLOGY	*Ophiocordyceps* is a genus of fungi named in 1931; *camponoti-rufipedis* is named after *Camponotus rufipes*, the species of carpenter ant on which the fungus was discovered.
CLASSIFICATION	Fungi • Ascomycota • Sordariomycetes • Hypocreales • Ophiocordycipitaceae

ANTONIO MELIC

Lilian's Widow Spider

FANGS FOR THE MEMORIES

Much like rumors of Mark Twain's death, the threat of spiders to humans has been greatly exaggerated and most stories of deadly spiders are more myth than fact. But among notorious arachnids, none are more emblematic than widow spiders of the genus *Latrodectus*. Although death from a widow spider bite is rare, their fierce reputation is nonetheless deserved. Females of some species cannibalize the much smaller males after mating, causing the origin of the common name "widow." Widow spiders can deliver a painful dose of the neurotoxin latrotoxin. The severity of such bites depends on the victim's body mass or weight, the amount of venom delivered, and how quickly medical care is administered. Recently discovered in Spain, *Latrodectus lilianae* is beautifully patterned and the latest of more than thirty widow spiders found on every continent except Antarctica. With this species being found in Europe, one can only imagine that many others await discovery in the biologically diverse regions of Asia, Africa, and South America.

Latrodectus lilianae male (left) and larger female (right) on web.

ANTONIO MELIC

DISCOVERED	Spain
SCIENTIFIC NAME	*Latrodectus lilianae* Melic 2000
ETYMOLOGY	*Latrodectus* is a genus of spiders named in 1805; *lilianae* named in dedication to Antonio Melic's wife, Lilian.
CLASSIFICATION	Animalia • Arthropoda • Arachnida • Aranae • Theridiidae

© W. WÜSTER

Ashe's Cobra

THE SPITTING IMAGE OF DANGER

© W. WÜSTER

Commonly known as cobras, snake species in the genus *Naja* are extremely venomous and quite dangerous to humans. While most cobras deliver venom by biting their victims, spitting cobras have developed a mechanism by which they can squirt venom almost ten feet. If the snake's toxin enters the eyes or touches the skin, blindness or tissue damage severe enough to require amputation of the affected limb can sometimes occur. At the time of its discovery, *Naja ashei* was the largest known spitting cobra. It averages a very scary seven feet in length, with some specimens as long as nine feet. Because this new species is so large, it can produce its venom in record-breaking quantities, making its attacks especially dangerous. Luckily, the venom of *N. ashei* seems to affect only the area where the spit lands and the poison doesn't appear to spread throughout the victim's body.

DISCOVERED	Kenya
SCIENTIFIC NAME	*Naja ashei* Wüster & Broadley 2007
ETYMOLOGY	*Naja* is a genus of venomous snakes named in 1768; *ashei* is in memory of James Ashe, for his contributions to East African herpetology.
CLASSIFICATION	Animalia • Chordata • Reptilia • Squamata • Elapidae

ALASTAIR S. ROBINSON

Attenborough's Pitcher

FEED ME, SEYMOUR

Carnivorous plants are primarily insect eating, but this new species of pitcher plant is so large, it could conceivably lure rodents or reptiles into its football-sized cup that holds the sugary fluid that attracts its usual prey. Discovered in 2007 in the Philippines, *Nepenthes attenboroughii* is one of the world's largest known pitcher plants, with a height that can reach five feet and a lower pitcher measuring about one foot tall by six inches wide. This strikingly beautiful new species was originally discovered in a single area at an elevation of 5,600 feet on Mount Victoria's summit, but has since been found on nearby peaks. Given its extremely limited range and declining numbers as a result of illegal poaching, this new species is believed to be critically endangered. Unlike Audrey II in *Little Shop of Horrors*, *N. attenboroughii* is definitely insectivorous and not known to *really* eat rats.

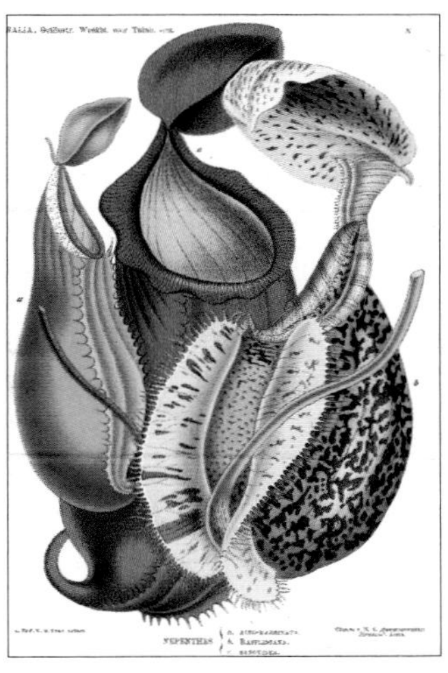

Nepenthes, pitcher plants, from "Stoomdrukkerim Floria" (1900).

COURTESY OF VINTAGE PRINTABLE

DISCOVERED	Mount Victoria, Palawan, Philippines
SCIENTIFIC NAME	*Nepenthes attenboroughii* Robinson, McPherson & Heinrich 2009
ETYMOLOGY	*Nepenthes* is a genus of carnivorous plants named in 1753; *attenboroughii* is named in honor of Sir David Attenborough, a naturalist, broadcaster, and patron of Philippine conservation efforts.
CLASSIFICATION	Plantae • Angiosperms • Eudicots • Caryophyllales • Nepenthaceae

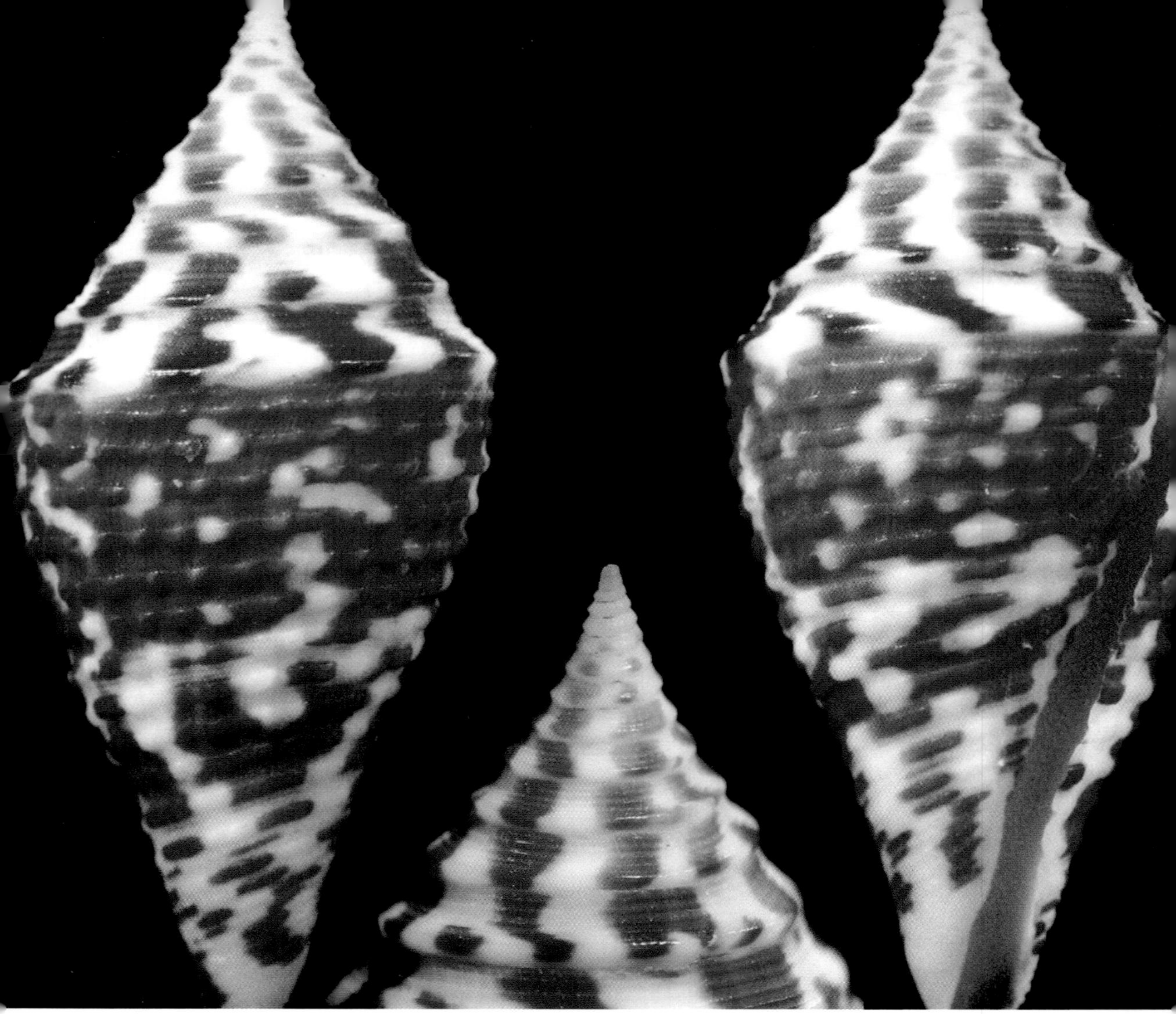

BALDOMERO M. OLIVERA

Andre Menez's Cone Snail

CONE OF (DEADLY) SILENCE

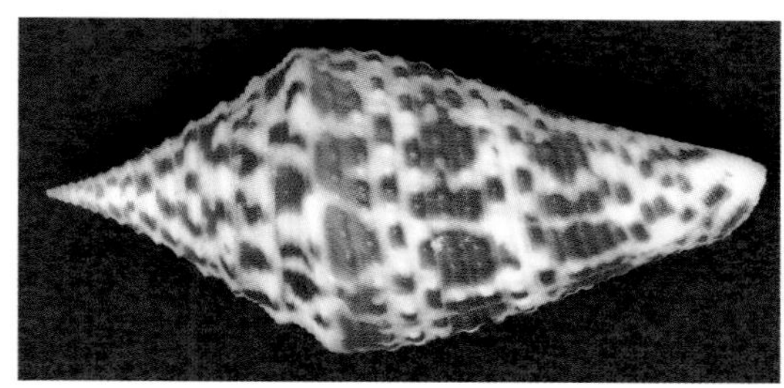

BALDOMERO M. OLIVERA

Between 2000 and 2010, approximately eighty new living species of cone snails were described, including *Conus andremenezi*, which was collected in the nets of trawl fishermen near Aliguay Island, Philippines. With more than six hundred species, cone snails are both diverse and one of the most venomous marine groups. In nature, cone snails use their teeth like harpoons to paralyze prey with a battery of toxins. The geographic cone snail (*C. geographus*) has such a fast-acting and powerful toxin that it has earned the nickname "cigarette snail"—after being stung by *C. geographus*, a human would have only enough time to smoke a cigarette before dying. Although the venom of cone snails—conotoxins—can be strong enough to kill a human, pharmaceutical research is continuing to find medical uses for them, such as nonaddictive painkillers and treatments for epilepsy and other neurological diseases.

DISCOVERED	Aliguay Island, Philippines
SCIENTIFIC NAME	*Conus andremenezi* Biggs, Watkins, Corneli, & Olivera 2010
ETYMOLOGY	*Conus* is a genus of mollusks named by Linnaeus in 1758; *andremenezi* is in honor of Andre Menez, for his contributions to the field of toxicology.
CLASSIFICATION	Animalia • Mollusca • Gastropoda • Neogastropoda • Conidae

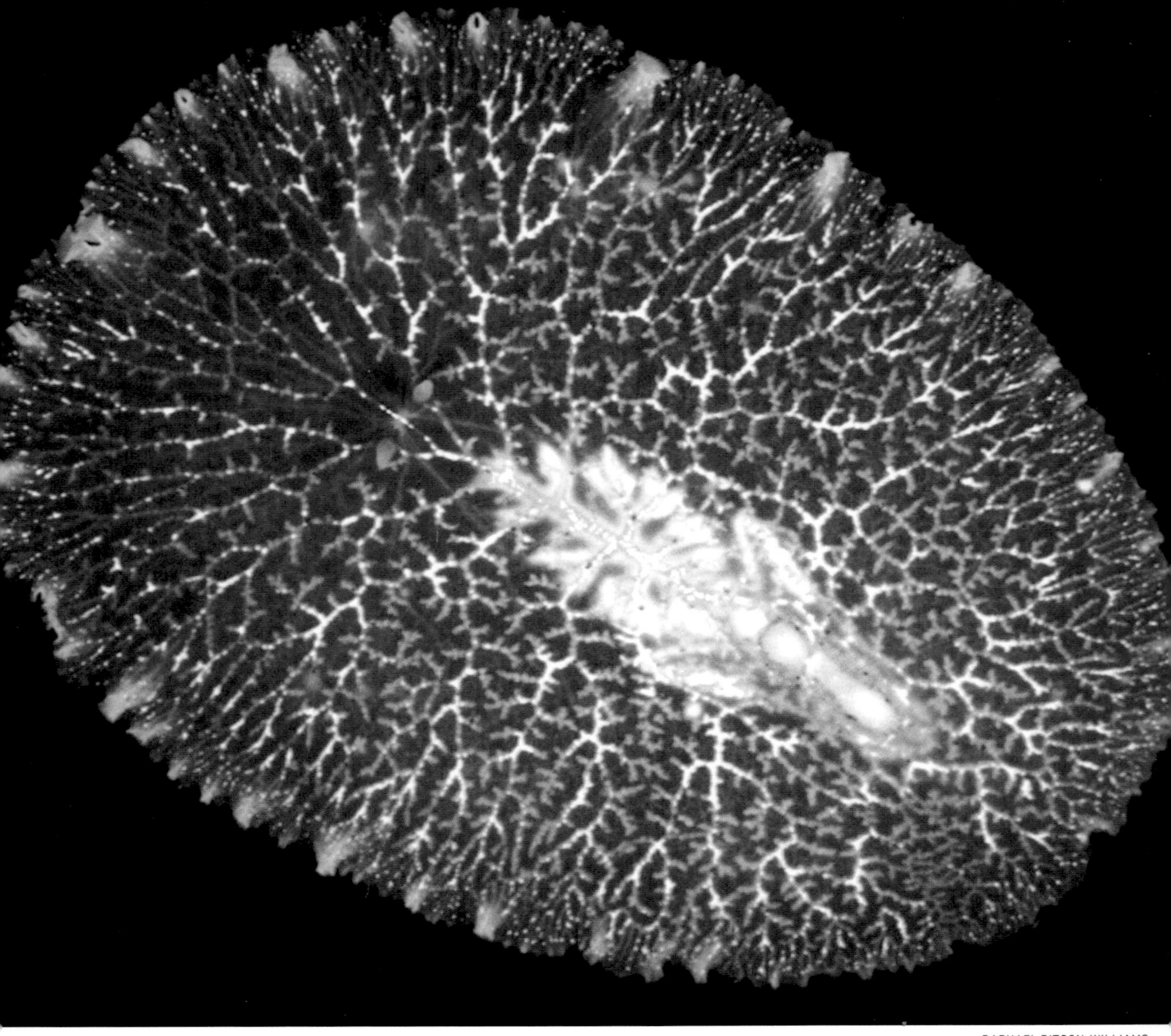

RAPHAEL RITSON-WILLIAMS

Poisonous Predaceous Polyclad

FROM FLATWORM TO FAT WORM: GONE IN 1,380 SECONDS

Ten times more poisonous than cyanide, tetrodotoxin is a powerful neurotoxin that was first isolated from puffer fish in the 1960s and has killed hundreds of people over the past sixty years. The evolution of the deadly tetrodotoxin as a defense against predators has been long recognized. However, a newly discovered and not yet named species of flatworm, "Planocerid species 1," has been shown to use the lethal tetrodotoxin as a weapon. Within twenty-three minutes of Planocerid species 1 being placed near a marine mollusk, the flatworm will wrap its body around the mollusk's shell, subdue it with the toxin, remove the paralyzed body, and eat it. The mollusk's defensive trapdoor is powerless against the attack, regardless of its size. Like many colorful polyclad flatworms, this new Planocerid is sometimes confused with the unrelated nudibranchs, which also inhabit coral reefs.

From flat to fat to flat again: flatworm eating a red-bodied cowry leaving just the cowry shell.

RAPHAEL RITSON-WILLIAMS

DISCOVERED	off the Pacific island of Guam
SCIENTIFIC NAME	"Planocerid species 1" Ritson-Williams, Yotsu-Yamashita & Paul 2006
ETYMOLOGY	as yet to be described
CLASSIFICATION	Animalia • Platyhelminthes • Rhabditophora • Polycladida • Planoceridae

LISA-ANN GERSHWIN

King's Deadly Jelly

TOXIN IN A BOX

While we may associate the oceans' most lethal predators with large, *Jaws*-like size, some of the smallest and most beautiful sea creatures—such as the box jellyfish—are just as harmful and deadly as the over-sized and overhyped bad boys. Box jellyfish are not fish at all but belong to the Cubozoa class of animals in the phylum Cnidaria. The venom of box jellies is one of the deadliest toxins in the world and stings from them have resulted in more than sixty documented human deaths. With a bell (box) size of only 1.2 inches in height, *Malo kingi* is a tiny jelly believed to have killed an American tourist in 2002 when he developed cubozoan Irukandji syndrome after brushing against a box jelly. *M. kingi* is difficult to see—not only is it tiny, it is also transparent. Even its sting may go unnoticed since it's reported to be fairly non-painful. Unfortunately, by the time symptoms begin to show it may be too late to start life-saving treatment; within twenty minutes of being stung, victims may die of cardiac arrest.

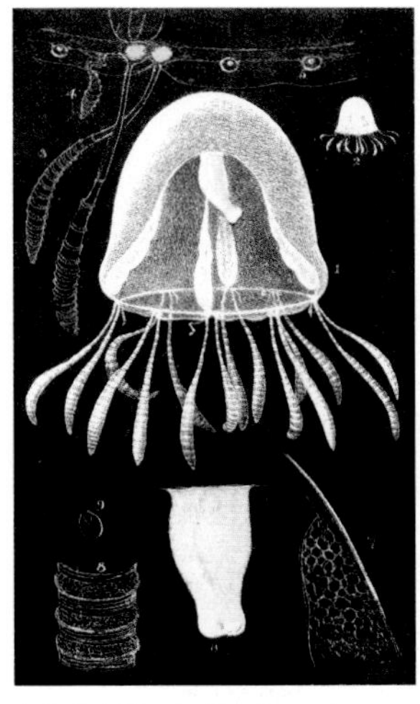

Jellyfish illustration from *A Naturalist's Rambles on the Devonshire Coast* (1853).

COURTESY OF VINTAGE PRINTABLE

DISCOVERED	far northern Queensland, Australia
SCIENTIFIC NAME	*Malo kingi* Gershwin 2007
ETYMOLOGY	*Malo* is a genus of jellyfish named in 2005; *kingi* named in honor of Robert W. King, who is believed to have died from being stung by this species.
CLASSIFICATION	Animalia • Cnidaria • Cubozoa • Carybdeida • Tamoyidae

- **A** Cave Pseudoscorpions
- **B** Morafka's Desert Tortoise
- **C** Yellowstone Bacterium
- **D** Nature Conservancy Diving Beetle
- **E** Cryptic Forest-Falcon
- **F** Stephenson's Antarctic Flower
- **G** Nepalese Autumn Poppy
- **H** Bare-Faced Bulbul
- **I** Siau Island Tarsier
- **J** Heat-Loving Tonguefish

Chapter 8

GOING TO EXTREMES

NEW SPECIES FROM THE MOST EXTREME ENVIRONMENTS

When considering the species that have recently been discovered in extreme environments, we're reminded of a line from *Jurassic Park*: "Life finds a way." And indeed it does, when you examine the range of conditions under which species can be found. From hot springs in Yellowstone to cold Alaskan waters; from intense sunlight in the Namib Desert to eternally dark trenches on ocean floors; from mountaintops to deep caves—life finds a way of unexpectedly existing and thriving in incredibly diverse and sometimes hostile environments. The ability of natural selection to wedge life forms into every nook and cranny of our planet in geologically and climatically extreme environments is nothing short of astounding. Life occupies just about every niche we can imagine and probably others we don't yet know.

When life finds a way to survive in an isolated or extreme environment, it is not surprising that new species arise. Small populations of organisms that remain physically and genetically isolated from their parental stock eventually evolve to become a separate species. Isolated populations may arise through a variety of factors: through dispersal, such as a plant seed reaching a remote island; through geographical changes, as when a barrier (such as a volcanic mountain) arises to divide a once-contiguous population; or as a result of climate change, which can lead to shifts in patterns of precipitation and temperatures fragmenting ecosystems. Strong selection pressures, like those of extreme environments, can accelerate the divergence of such isolated populations.

By studying species in extreme environments, we learn some of the most compelling stories of evolutionary adaptation. Certain characteristics are repeatedly associated with particular kinds of extreme habitats: trees at high altitudes are dwarfed, insects found in caves are blind and have disproportionately elongated legs and antennae, and animals such as birds and insects can become flightless on small, isolated islands. Properly adapted organisms can survive remarkably well in extreme conditions. One of the species we include in this chapter is an aquatic beetle that lives in subterranean aquifers (underground layers of water). Among its adaptations to life underground are the loss of its eyes, wings, and coloration. Until a few years ago, no one suspected that any insect would be found in an aquifer. We now know that there are hundreds of species of aquifer beetles on at least two continents that have adapted to this unusual life, literally out of sight.

Dramatic examples of extreme habitats include the so-called black smokers and white smokers, the deep-sea vents that support entire, alien-seeming ecosystems. These hydrothermal sea vents are typically found near volcanic activity on the seafloor where their plumes emit superheated steam and chemicals. While most species are parts of ecosystems that derive energy from the sun through photosynthesis, these sea-vent communities live in the extreme depths of the oceans where sunlight cannot penetrate. These remarkably complex ecosystems are based instead on the chemicals that are belched from these geothermal vents and include species from bacteria to fish.

Astrobiologists (sometimes called exobiologists) hope to discover life on other planets. In order to explore the limits of conditions under which life might arise, they focus on the most extreme habitats on Earth. As a result, the boundary conditions for life continue to be redefined. The devil's worm (*Halicephalobus mephisto*) was discovered in 2011 nearly a mile beneath the Earth's surface in one of the world's deepest mines. This roundworm, or nematode, is the first multicellular organism known to survive at such depth, pressure, and temperature. Water from the borehole in which it was found had not been in contact with the atmosphere for the last four thousand to six thousand years. Its discovery has implications for the possibility of "higher" life forms beneath the surface of other planets.

Recalling that 80 percent of Earth's "higher" plants and animals are unknown to us and that we have not yet discovered 99 percent of the microbes, it is safe to say that shocking discoveries of life forms under extreme conditions will continue to be made for generations to come. Our planet is dynamic and constantly changing. For billions of years, life has responded to such change by adapting, again and again, in order to survive, thrive, and multiply. To surprising extremes, life finds a way. As we continue to explore the limits of the biosphere, we will be rewarded with the discoveries of many new species and the remarkable ways in which they have adapted to life on Earth.

COURTESY OF VINTAGE PRINTABLE

Titanobochica magna.

ANA SOFIA REBOLEIRA, ZARAGOZA & REBOLEIRA 2010

Cave Pseudoscorpions

SPECTACULAR SCORPIONS, TRUE OR FALSE?

Parobisium yosemite, new species.

DR. JEAN K. KREJCA, ZARA ENVIRONMENTAL LLC

Pseudoscorpions are small, eight-legged animals with pear-shaped bodies and pincers that look so much like scorpions that people may jump and scream—if they even see them. Almost all false scorpions are harmless, although a recently discovered new Colorado cave species, *Cryptogreagris steinmanni*, was found to have venom-tipped pincers. Most pseudoscorpions are overlooked due to their tiny size and secretive habits, such as living under stones or in tree holes. Discovered in four Portuguese limestone caves, new species *Titanobochica magna* is notable for its unusually large size (about one-quarter of an inch) and for the curious separation of living quarters between the adults and nymphs (the immature preadults). Like human parents who want to get away from their teenagers, the adults were found in the deepest, most isolated parts of the caves, while the kids hung out in areas closer to the caves' surfaces where there was more organic matter—also known as pseudoscorpion food. Discovered in 2006, *Parobisium yosemite* is another new cave-inhabiting species that was found in two Yosemite National Park granite caves. While the Yosemite caves are only about two hundred years old, pseudoscorpions have been around for a long time, with fossils dating to about 380 million years ago.

DISCOVERED	southern Portugal
SCIENTIFIC NAME	*Titanobochica magna* Reboleira, Zaragoza, Gonçalves & Oromí 2010
ETYMOLOGY	*Titanobochica* is from the Latin word *titan*, meaning giant, and the name Bochia, an important god in ancient Colombia; *magna* is from the Latin *magnus*, meaning big, to reflect the species' large size.
CLASSIFICATION	Animalia • Arthropoda • Arachnida • Pseudoscorpionida • Bochicidae
ENVIRONMENT	deep cave

TAYLOR EDWARDS

Morafka's Desert Tortoise

BUT IT'S A DRY HEAT

TAYLOR EDWARDS

Distinguishing among species of land tortoises of the genus *Gopherus* can be challenging. For example, new species *Gopherus morafkai* simply varies from other *Gopherus* species by having relatively smaller front feet and only modest differences in the shape of its shell. This species story, however, is a taxonomic tale that begins in 1860 when a U.S. Army doctor found a little, palm-sized tortoise while stationed in the Arizona Territory. A year later, the tortoise was presented to the California Academy of Sciences as new species *Xerobates agassizii* (but later became known as *G. agassizii*). By the early 2000s, scientists began to suspect that specimens of *G. agassizii* could actually be more than one species. Using DNA analysis, it was determined that *G. agassizii* was at least two distinct tortoises with two different habitats: *G. agassizii*, which lives north and west of the Colorado River, and new species *G. morafkai*, which lives in the extreme environment of the Sonoran Desert in southern Arizona—and in the reptile zoo of Life Sciences Building A at Arizona State University.

DISCOVERED	Arizona, USA
SCIENTIFIC NAME	*Gopherus morafkai* Murphy, Berry, Edward, Leviton, Lathrop & Riedle 2011
ETYMOLOGY	*Gopherus* is a genus of tortoises named in 1832; *morafkai* is named in honor of Professor David Joseph Morafka in recognition of his contributions to *Gopherus* biology and conservation efforts.
CLASSIFICATION	Animalia • Chordata • Reptilia • Testudines • Testudinidae
ENVIRONMENT	desert

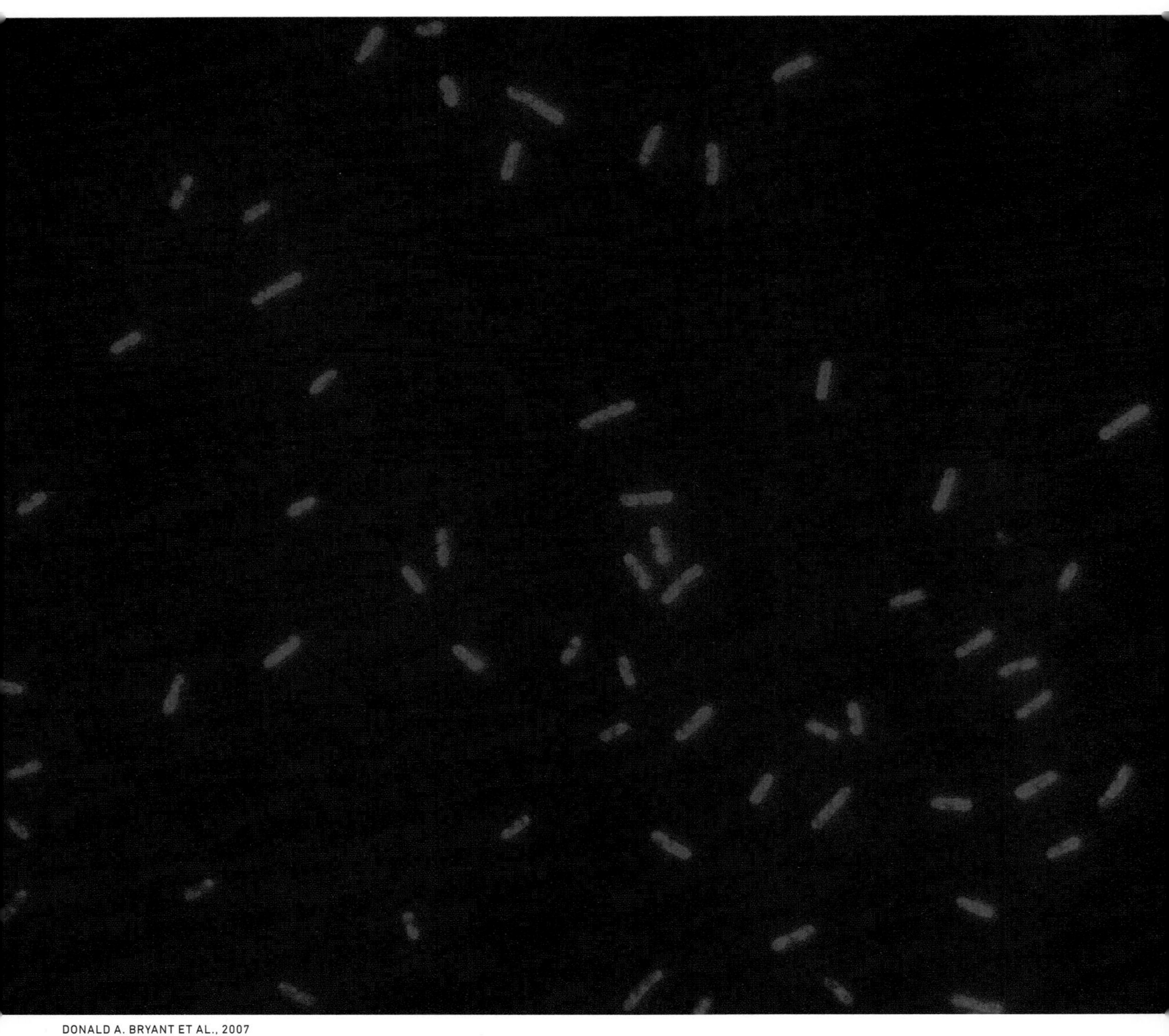

DONALD A. BRYANT ET AL., 2007

Yellowstone Bacterium

TRIPPING THE LIGHT FANTASTIC

With the recent discovery of *Candidatus* Chloracidobacterium thermophilum in three hot springs of Yellowstone National Park, scientists have found a new bacterium that is capable of harvesting energy from sunlight through a novel form of photosynthesis (the process of converting light energy to chemical energy). Living in alkaline water that reaches temperatures of 122 to 151 degrees Fahrenheit, these bacteria are members of microbe communities that form the drifting mats of color for which the hot springs are known. Yellowstone's hot springs may be the world's most important biodiversity hot spot for heat-loving microbes and, despite being studied for more than fifty years, new bacteria species continue to be discovered in these extreme environments. As humans deplete Earth's fossil fuels, energy-producing bacteria like *C. C.* thermophilum have amazing importance as a potential source of biofuel.

Locale of *C. thermophilum*, Octopus Spring, Yellowstone National Park, Wyoming, USA.

DR. DAVID M. WARD

DISCOVERED Wyoming, USA

SCIENTIFIC NAME *Candidatus* Chloracidobacterium thermophilum Bryant, Garcia Costas, Maresca, Chew, Klatt, Bateson, Tallon, Hostetler, Nelson, Heidelberg & Ward 2007

ETYMOLOGY *Chloracidobacterium* is based on a combination of the Greek word *chloro*, meaning green, and the Latin words *acido*, meaning acid, and *bacterium*, meaning "small rod"; *thermophilum* is based on the Greek words *thermo*, meaning heat, and *phylum*, meaning love. Chlorophyll allows organisms to produce energy from light. Thus, the species name means an "acid- and heat-loving bacterium that produces energy from light."

CLASSIFICATION Bacteria • Acidobacteria • Acidobacteria • Acidobacteriales • Acidobacteriaceae

ENVIRONMENT chemicals and heat

KELLY B. MILLER

Nature Conservancy Diving Beetle

WELL-GROUNDED IN TEXAS

For many years, aquatic beetle species living in the subterranean waters of caves or in groundwater aquifers were believed to be extremely rare. However, work in Australia conducted in 2007 and 2008 revealed at least one hundred new species of aquatic subterranean or groundwater-adapted diving beetles. As of 2009, subterranean diving beetles found in North America represented four genera, including a new genus and species, *Ereboporus naturaconservatus*. Using drift nets over the openings of Caroline Springs at the headwaters of Independence Creek in Texas, biologists collected *E. naturaconservatus* in 2007 and 2008. The openings where the beetle was discovered emerge from the Edwards-Trinity Aquifer, which is part of an expansive subterranean water system that covers 77,220 square miles—over (and under) one-fourth of Texas.

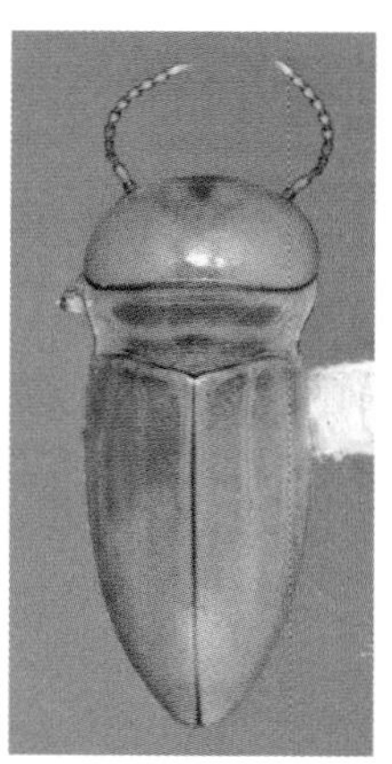

KELLY B. MILLER

DISCOVERED	Terrell County, Texas, USA
SCIENTIFIC NAME	*Ereboporus naturaconservatus* Miller, Gibson & Alarie 2009
ETYMOLOGY	*Ereboporus* is from the Greek word *erebos*, meaning hell, to reflect the depths at which the species was found, and *-porus*, a common root for genus names in the group; *naturaconservatus* is from the Latin words *natura* and *conservatus*, meaning nature and conservation to honor the Nature Conservancy and its work to preserve Independence Creek.
CLASSIFICATION	Animalia • Arthropoda • Insecta • Coleoptera • Dytiscidae
ENVIRONMENT	aquifer

ANDREW WHITTAKER

Cryptic Forest-Falcon (Falcão Cryptico)

AMAZONIAN AMBIGUITY

ANDREW WHITTAKER

Sometimes new species are discovered when they're first heard rather than seen—such as a new monkey species in Myanmar (*Rhinopithecus strykeri*) that was first detected because it sneezed in the rain, or the Pernambuco Pygmy-Owl (chapter 3) that was discovered because it gave a hoot. Similarly, the Cryptic Forest-Falcon (*Micrastur mintoni*) was discovered because its distinctive call was heard before it was even spotted. "Cryptic" species are two or more closely related species that are difficult to distinguish on sight but are, nonetheless, distinct and separate kinds of animals or plants. Not only is the Cryptic Forest-Falcon's call distinctive, but it has a single white tail band in addition to a narrow white tail tip. For almost a century, museum specimens of *M. mintoni* had been misidentified as *M. gilvicollis*, another forest falcon, which it strongly resembles with the exception of its tail feathers.

DISCOVERED	Brazil
SCIENTIFIC NAME	*Micrastur mintoni* Whittaker 2002
ETYMOLOGY	*Micrastur* is a genus of forest falcons named in 1841; *mintoni* is named in honor of Clive D. T. Minton, for his contributions to shorebird biology and his work in their worldwide conservation.
CLASSIFICATION	Animalia • Chordata • Aves • Falconiformes • Falconidae
ENVIRONMENT	rain forest

COURTESY OF P. J. LOPEZ-GONZALEZ (UNIVERSIDAD DE SEVILLA, SPAIN)

Stephenson's Antarctic Flower

FROZEN CARROTS IN A BANANA BELT

COURTESY OF P. J. LOPEZ-GONZALEZ (UNIVERSIDAD DE SEVILLA, SPAIN)

When we think about the animals that live at the top and bottom of our planet, most of us tend to visualize polar bears and penguins. While the polar regions may seem like inhospitable habitats, there are actually thousands of species that survive the extreme environments and subfreezing temperatures of the Arctic and Antarctic Circles. In 2008, an Antarctica marine survey resulted in the collection of thirty thousand specimens, including a new sea anemone species, *Stephanthus antarcticus*. Found at depths of up to four thousand feet off the South Shetland Islands, the carrotlike appearance of this new anemone is so unusual that it had to be assigned to a new genus. Its habitat is called "Antarctica's Banana Belt" even though 80 to 90 percent of the islands are permanently covered in glaciers and summer temperatures reach a balmy 32 to 34 degrees Fahrenheit.

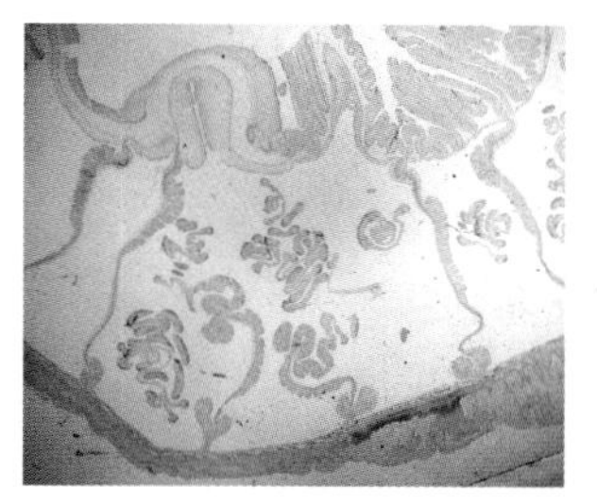

Internal anatomy of *Stephanthus antarcticus*.

E. RODRIGUEZ (AMNH, NY)

DISCOVERED	South Shetland Islands, Antarctica
SCIENTIFIC NAME	*Stephanthus antarcticus* Rodriguez & Lopez-Gonzalez 2003
ETYMOLOGY	*Stephanthus* is a combination of *Steph*, in honor of Professor T. A. Stephenson, and *anthus* after the Greek word *anthos*, meaning flower, commonly used for sea anemone names; *antarcticus* is used to reflect Antarctica, where the species was found.
CLASSIFICATION	Animalia • Cnidaria • Anthozoa • Actiniaria • Hormathiidae
ENVIRONMENT	polar regions

PAUL EGAN

Nepalese Autumn Poppy

ELEVATED BEAUTY

Growing on the steep terrain of the Himalayan Mountains, *Meconopsis autumnalis* is a spectacular new species of poppy. Standing over five feet tall, with beautifully brilliant yellow flowers, it may be surprising that this poppy escaped the attention of science until now. No doubt this is largely due to the species' remote location and a habitat with an elevation of over 10,000 to almost 14,000 feet. Specimens of *M. autumnalis* have actually been collected twice before—first in 1962 by the storied Himalayan plant hunter Adam Stainton, and again in 1994 by members of the University of Tokyo's Department of Plant Resources. Yet these specimens were never described or studied in sufficient detail to reveal their new species status. The expeditions that rediscovered the poppy in 2008 had to be completely self-sufficient as they worked during the heavy rains of the autumn monsoon season and miles from the nearest human habitation. *M. autumnalis* is locally abundant and blooms from late July through September.

PAUL EGAN

DISCOVERED	Himalayas, Nepal
SCIENTIFIC NAME	*Meconopsis autumnalis* Egan 2011
ETYMOLOGY	*Meconopsis* is a genus of poppies named in 1814; *autumnalis* to reflect the autumn season, when the plant flowers.
CLASSIFICATION	Plantae • Magnoliophyta • Magnoliopsida • Ranunculales • Papaveraceae
ENVIRONMENT	high elevation

IAIN WOXVOLD

Bare-Faced Bulbul

BALD BUT BUBBLY

IAIN WOXVOLD

On average, about eighteen thousand new species are described every year, but of those, only seven or eight are new bird species—less than 0.04 percent! It's no surprise, then, that it's taken more than a hundred years for the first new Asian songbird species to be found. Discovered in a limestone karst region of Laos, new species *Pycnonotus hualon* is distinctive for its bald face and the whistled, dry bubbling notes of its song. Karst environments are notable for their lack of surface water and their rock formations, sinkholes, and caves. According to the scientists who discovered this distinctive bird, *P. hualon* may represent an extreme example of a passerine songbird (the family Pycnonotidae) that can be found only in a specific habitat and is limited to the limestone belt in central Indochina.

DISCOVERED	Lao People's Democratic Republic (Laos)
SCIENTIFIC NAME	*Pycnonotus hualon* Woxvold, Duckworth & Timmins 2009
ETYMOLOGY	*Pycnonotus* is a genus of songbirds named in 1826; *hualon* from the Lao word meaning "bald headed" to reflect the bird's somewhat bare and featherless head.
CLASSIFICATION	Animalia • Chordata • Aves • Passeriformes • Pycnonotidae
ENVIRONMENT	dry limestone karst

GEOFF DEEHAN

Siau Island Tarsier

PLEASED TO MEAT YOU

Small enough to sit in your hand, tarsiers are big-eyed, long-tailed primates that are nocturnal, arboreal, and the only living primate known to be completely carnivorous—no veggies for these guys. Unlike other closely related tarsiers, new species *Tarsius tumpara* is distinctive for its call, hair color, and the distribution of the fur on its body and tail. *T. tumpara* was first captured in 2002 on Siau, a tiny, remote Southeast Asia island dominated by Mount Karengetang, which makes up 55 percent of the island's landmass. The vast majority of Siau's population—including fifty thousand humans and *T. tumpara*—is crowded onto the island's southernmost tip to avoid the dangers of Mount Karengetang, one of the most active and dangerous volcanoes in Indonesia. Unfortunately, these same humans may be hunting *T. tumpara* to extinction.

Mount Karengetang on Siau Island, habitat of *Tarsius tumpara*.

GEOFF DEEHAN

DISCOVERED	Siau Island, North Sulawesi
SCIENTIFIC NAME	*Tarsius tumpara* Shekelle, Groves, Merker & Supriatna 2008
ETYMOLOGY	*Tarsius* is a genus of primates named in 1780; *tumpara* comes from the word the inhabitants of Siau Island use for it.
CLASSIFICATION	Animalia • Chordata • Mammalia • Primates • Tarsiidae
ENVIRONMENT	active volcano, remote island

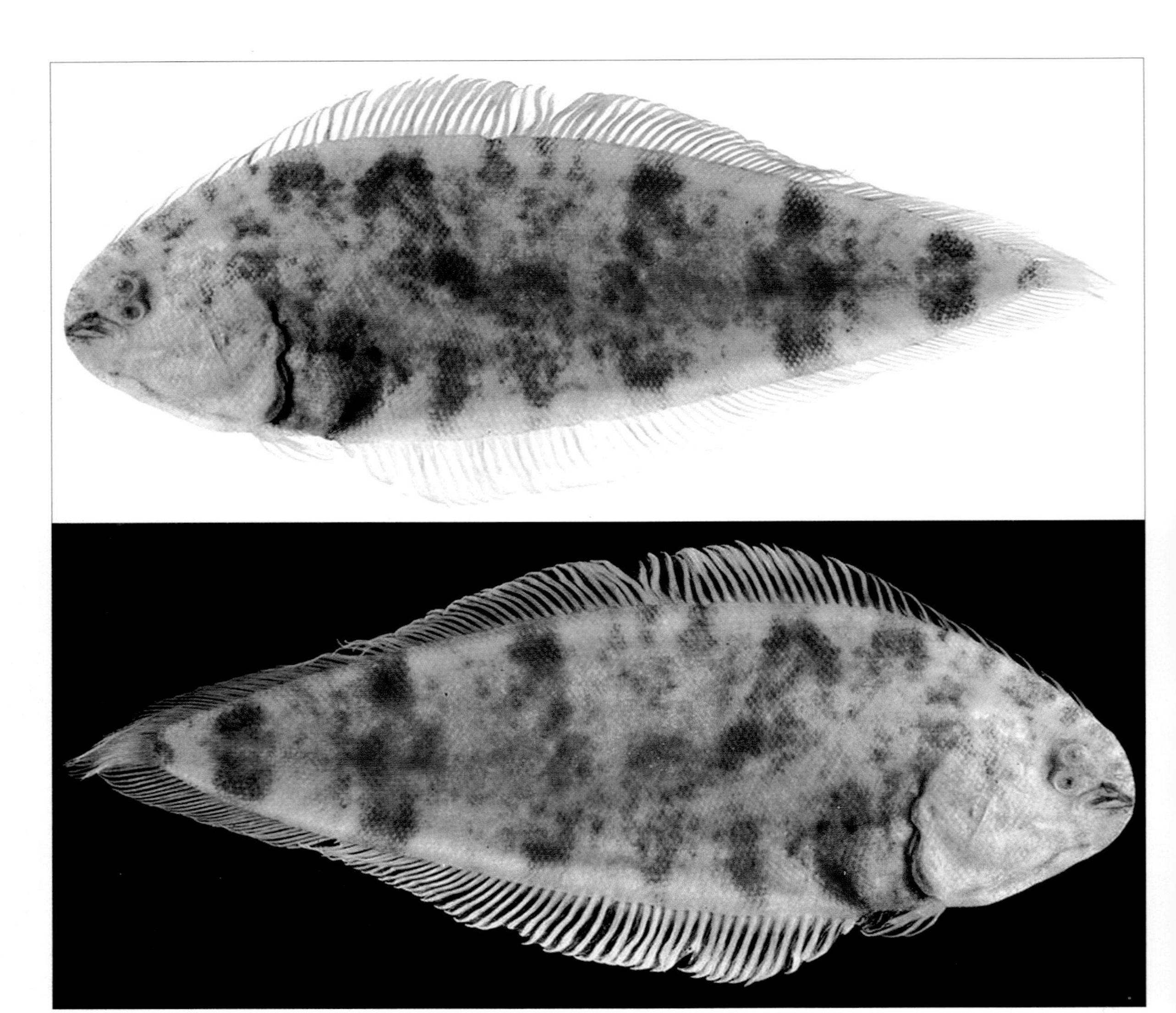

THOMAS A. MUNROE, NMFS / NOAA

Heat-Loving Tonguefish

SMOKIN' HOT HABITAT

Often called flatfish, Pleuronectiformes are a type of rayfish with more than four hundred species worldwide that have characteristically flat bodies and protruding eyes on only one side of their head, giving them an asymmetrical, googly-eyed appearance. New tonguefish species *Symphurus thermophilus* is a tiny flatfish that measures no more than four inches snout to vent, with an unusually deep body for a "flat" fish and eyes that are bigger on average than other tonguefish species. *S. thermophilus* is also the first Pleuronectiform known to live in the extreme environment of an active "white smoker" hydrothermal sea vent at depths of 3,280 feet and temperatures of 356 degrees Fahrenheit. Living in this extreme environment appears to have produced some oddities that may be due to the superheated or chemically charged water—many of the specimens had fused skeletons while others had deformed or missing fins. Great googly moogly.

Symphurus thermophilus in situ (as found in nature).

THOMAS A. MUNROE, NMFS/NOAA

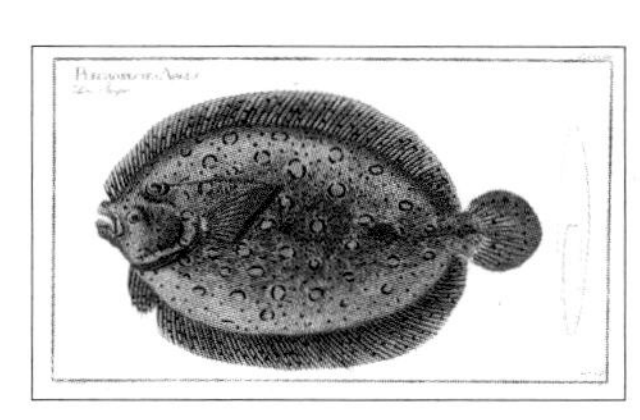

Colored engraving of the Argus flounder by Gabriel Bodenehr (c. 1785–1799).

COURTESY OF VINTAGE PRINTABLE

DISCOVERED	Kaikata Seamount off southern Japan in the Pacific Ocean
SCIENTIFIC NAME	*Symphurus thermophilus* Munroe & Hashimoto 2008
ETYMOLOGY	*Symphurus* is a genus of tonguefish named in 1810; *thermophilus* is in reference to the species' habitat of hydrothermal vents and is based on the Greek words *thermos*, meaning heat, and *philos*, meaning lover.
CLASSIFICATION	Animalia • Chordata • Actinopterygii • Pleuronectiformes • Cynoglossidae
ENVIRONMENT	active hydrothermal vent (a white smoker)

- A Groves's Nudibranch
- B Firefly Flasher
- C Appalachian Tiger Swallowtail
- D Mache Mountains Glass Frog
- E Fish Mime
- F Ngome Dwarf Chameleon
- G Shocking Pink Millipede
- H Oria's Leaf Insect
- I Denise's Pygmy Seahorse
- J Jumpin' Spider Ant-Mimic

Chapter 9

THE HIGHEST FORM OF FLATTERY

THE BEST NEW SPECIES MIMICS

To paraphrase an old adage, "Don't judge a species by its appearance." There is a vast and growing list of species that resemble something they are not, and people, as well as predators, are frequently fooled. Before we discuss some types of mimicry, it is worth mentioning two extremes of species appearance: animals or plants that have evolved to blend in with their environments, and those that unmistakably stand out. Stick insects and chameleons are familiar examples of *camouflage*, in which animals adapt to match their surroundings and hide from predators. It is an incredibly impressive type of mimicry when done well. So, too, is *aposematism*, in which a species displays a bright and easily recognized color to convey a warning to predators that the species is poisonous or bad tasting. Skunks and poison dart frogs provide two common examples of aposematism, but many other animal species, such as millipedes and octopuses, use this mimicry technique.

Perhaps the best-known types of mimicry are those described as *Batesian* and *Müllerian*. In Batesian mimicry, a harmless or tasty species will model its appearance after a poisonous or dangerous species to increase the likelihood that a predator will leave it alone. For example, the beautiful red-, black-, and yellow-ringed milk snake is harmless but resembles the similarly marked but venomous coral snake. You can even find examples of Batesian mimicry in your own backyard: if you look closely at the insects visiting your flower bed, you may see harmless flower flies with yellow and black stripes that mimic those of bees and wasps. Müllerian mimicry, on the other hand, involves groups of closely related species in which all are distasteful or dangerous. The species converge upon a common appearance so that if a predator feeds upon or encounters any one of them, it will subsequently avoid all of them. South American butterflies of the genus *Heliconius* are an example of Müllerian mimicry: all of the species share similar color patterns and all are toxic to predators.

Mimicry is not limited to cases of defense or warning. There are also many examples of aggressive mimicry, wherein a predator will mimic another species in order to attract or get close to prey. This type of mimicry is sometimes called *Peckhamian*. You now know about the deep-sea anglerfishes that wiggle wormlike appendages on their heads to bait their prey. Sometimes the "bait" is illuminated by pockets of bioluminescent bacteria that make it visible in the deep sea below the light zone. In this case, the anglerfish has mimicked a meal but, instead, other fish become a meal themselves. There are many other examples of mimicry that do not involve vision and instead are based on sound or scent. The spotted predatory katydid in Australia mimics the call of female cicadas. This attracts unsuspecting male cicadas, who instead of finding a mate find themselves becoming the katydid's dinner. Similarly, the bolas spider produces chemicals that imitate the sex pheromones of certain moth species and, once attracted to the scent of a possible mate, the moth is captured and consumed by the spider.

The roots of mimicry are not limited to the animal kingdom. While moths and lizards may look like the bark of a tree trunk or green-colored insects blend in with leaves, the opposite is sometimes true of plants. Friar's cowl and a cocklebur are two plants that have black spots that are the same size, shape, and color of ants. There are also several species of legumes whose seed pods have red spots that make them look like caterpillars. In each case, it has been suggested that this mimicry repels the animals that would eat the plant but would avoid the insect.

In this chapter, we share a sampling of recently discovered species that use shape, color, or behavior to blend in, stand out, or mimic something they are not. Even though chameleons are a virtual cliché of camouflage in nature, we could not resist including at least one forest dwarf that is especially endearing. Other choices will be a bit more surprising—from fish behaving badly to get a school lunch to spiders posing as ants. To us, this all adds up to an incredible demonstration of evolution, the endless inventiveness of natural selection, and a heightened appreciation for the maze of dangers that many species face in their struggle for survival.

COURTESY OF VINTAGE PRINTABLE

ALICIA HERMOSILLO

Groves's Nudibranch

SPONGY WANNABE

Discovered in the rocky reefs of shallow subtidal water, *Berthella grovesi* is one of those nudibranchs that is neither colorful nor strikingly beautiful. However, its brownish white appearance and oval body allow it to blend perfectly with the sponges that look just like it. Interestingly, nudibranchs are the primary predators of sponges, but sponges also produce chemicals as a defense mechanism to discourage other animals from killing them. In addition to eating these sponges, *B. grovesi* may, in fact, be using Batesian mimicry—perhaps by modeling itself after a poisonous sponge, it avoids being eaten as well. Across the sixteen other known species of *Berthella* nudibranchs that are found around the world, *B. grovesi* is distinctive for its tubercles, the protruding nodes that give this sea slug its bumpy, mottled appearance.

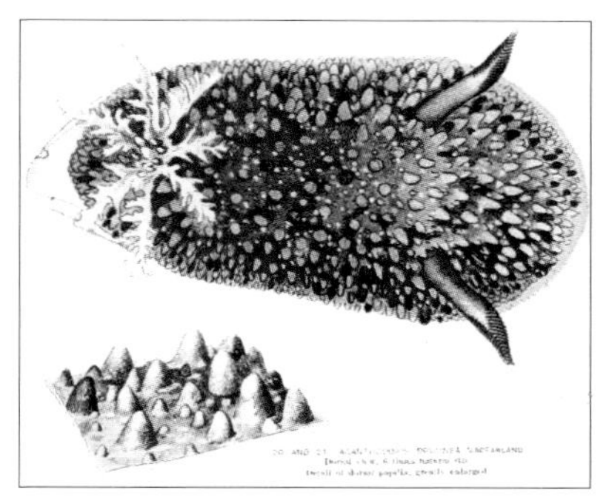

Brown sea slug, Bulletin of the Bureau of Fisheries (1905).

COURTESY OF VINTAGE PRINTABLE

DISCOVERED	Pacific coast of Mexico
SCIENTIFIC NAME	*Berthella grovesi* Hermosillo & Valdés 2008
ETYMOLOGY	*Berthella* is a genus of nudibranchs named in 1824; *grovesi* is named in recognition of Lindsey Groves, collections manager of malacology for the Natural History Museum of Los Angeles County.
CLASSIFICATION	Animalia • Mollusca • Gastropoda • Nudibranchia • Pleurobranchidae
TYPE OF MIMICRY	camouflage or Batesian

JAMES E. LLOYD

Firefly Flasher

FEMME FATALE FIREFLY

JAMES E. LLOYD

The bioluminescent light shows produced by fireflies on summer nights can be a beautiful sight for humans, but they can also be fatal to the firefly's prey. Some fireflies use light flashes from their light-emitting organs not only to attract mates of their own species but also to have other fireflies over for an evening meal. Although not unique in its aggressive mimicry among fireflies, females of new species *Photuris trivittata* are an illuminating example of how they mimic *behavior* rather than imitating general body appearance. The females of *P. trivittata* can produce multiple patterns of flashing, including one that they use in their own mating ritual and another that mimics the flashing pattern used by another firefly species. The result? Males of the other species are attracted to *P. trivittata* thinking they have found a mate, when, in fact, *P. trivittata* has found herself a dinner date.

DISCOVERED	Mexico
SCIENTIFIC NAME	*Photuris trivittata* Lloyd & Ballantyne 2003
ETYMOLOGY	*Photuris* is a genus of fireflies named in 1833; *trivittata* is named after the Latin words *tria*, meaning three, and *vittat*, meaning stripe.
CLASSIFICATION	Animalia • Arthropoda • Insecta • Coleoptera • Lampyridae
TYPE OF MIMICRY	aggressive

Top left to right: female (ventral), female black form (ventral).
Bottom left to right: female (dorsal), male (ventral).

RON GATRELLE ON BEHALF OF THE INTERNATIONAL LEPIDOPTERA SURVEY

Appalachian Tiger Swallowtail

TIGER OF A TAIL

First painting of an American butterfly (1587) by John White, expedition leader of Sir Walter Raleigh's Roanoke Island settlement, reproduced in Holland (1931).

COUTESY OF THE BRITISH MUSEUM

Native to North America, the eastern tiger swallowtail was the first New World butterfly to be depicted by European settlers more than four hundred years ago. In the 1800s, scientists began to notice that some butterflies use mimicry as a protection strategy from predators. A classic example is the viceroy butterfly, which imitates the color patterns of the vile-tasting monarch butterfly to avoid predators. (Monarchs are so distasteful that predators have been known to vomit after eating one!) Eastern tiger swallowtails are especially interesting because only the black females mimic another butterfly, called the pipevine swallowtail—so named because they eat the pipevine plant, which also makes them distasteful. During field studies conducted between 1985 and 2001, a new tiger swallowtail species, named *Pterourus appalachiensis*, was discovered in the Appalachian Mountains. In 2004, a black female was discovered, which suggests that the female of this species may be a mimic, too.

DISCOVERED	North Carolina, USA
SCIENTIFIC NAME	*Pterourus appalachiensis* Pavulaan & Wright 2002
ETYMOLOGY	*Pterourus* is a genus of butterflies named in 1777; *appalachiensis* is named after the Appalachian Mountains, where the species was discovered.
CLASSIFICATION	Animalia • Arthropoda • Insecta • Lepidoptera • Papilionidae
TYPE OF MIMICRY	Batesian

LUIS A. COLOMA

Mache Mountains Glass Frog

IT ISN'T EASY BEING GREEN

Translucent and typically green, frogs that belong to the family Centrolenidae can have skin with such complete transparency that bones and internal organs—such as intestines and beating hearts—can easily be seen from their ventral (under) side. And like a scene from the sci-fi movie *Predator*, glass frogs are so clear that they can virtually disappear in their leafy environments, enabling them to be almost invisible to both prey and predator. Despite their camouflage, nocturnal habits, and arboreal nature, several new species of glass frogs have been discovered this past decade. Found in Ecuador, *Cochranella mache* is one of the most beautiful glass frogs of the approximately 147 known species, found from Mexico to Argentina.

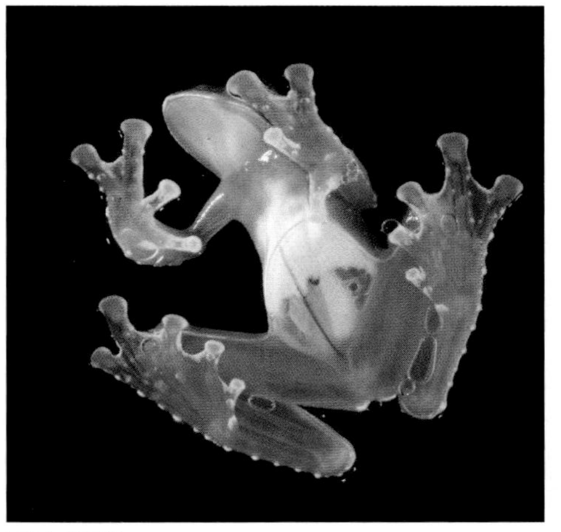

LUIS A. COLOMA

DISCOVERED	Riachuelo La Ducha in Esmeraldas Province, Ecuador
SCIENTIFIC NAME	*Cochranella mache* Guayasamin & Bonaccorso 2004
ETYMOLOGY	*Cochranella* is a genus of glass frogs named in 1951; *mache* after the Mache Mountains, where the species was found.
CLASSIFICATION	Animalia • Chordata • Amphibia • Anura • Centrolenidae
TYPE OF MIMICRY	camouflage

DR. M. HORI

Fish Mime

SCHOOL LUNCH BULLIES

Many species of fish live in social communities called shoals and swim in schools to increase their hunting ability, to find a mate, and to protect themselves against predators. These shoals usually include members of one species, but in certain cases, unrelated species come along for the ride. But what happens if the mimicking fish that joins your school is really your nemesis? When hunting prey, the coloration of new African species *Lepidiolamprologus mimicus* imitates the coloration of *Paracyprichromis brieni*. This mimicry allows *L. mimicus* to blend into a school of *P. brieni*, where it can then easily attack. *L. mimicus* is the first known example of aggressive mimicry in a fish genus that was described more than one hundred years ago.

L. mimicus (foreground), *P. brieni* (background).

DR. M. HORI

DISCOVERED	Lake Tanganyika, Zambia
SCIENTIFIC NAME	*Lepidiolamprologus mimicus* Schelly, Takahashi, Bills & Hori 2007
ETYMOLOGY	*Lepidiolamprologus* is a genus of cichlid fish named in 1903; *mimicus* is the Latin word meaning mimic, based on the Ancient Greek word *mimikos*, meaning "belonging to mimes," and is used to reflect the species' ability to imitate the coloring of its prey.
CLASSIFICATION	Animalia • Chordata • Actinopterygii • Perciformes • Cichlidae
TYPE OF MIMICRY	aggressive

COLIN TILBURY

Ngome Dwarf Chameleon

NGOME FOREST GNOME

Chameleon species have famously developed a special ability to mimic the colors of their immediate surroundings in order to avoid detection by predators. Found only in the South African Ngome Forest known as the "Mist Belt," *Bradypodion ngomeense* is a new species of dwarf chameleon discovered in an isolated area that is known for its diversity of chameleon species. This chameleon was first collected in 1992 but thought to be *B. transvaalense*, another dwarf chameleon known to have several different appearances—as chameleons do. Throughout this century, additional specimens of this charming chameleon with its angular casque (helmet) were found and analyzed. Using molecular methods as well as traditional taxonomic tools, scientists were able to determine that *B. ngomeense* was actually a new species.

Bradypodion ngomeense (top and bottom)

COLIN TILBURY

DISCOVERED	KwaZulu-Natal, South Africa
SCIENTIFIC NAME	*Bradypodion ngomeense* Tilbury & Tolley 2009
ETYMOLOGY	*Bradypodion* is a genus of chameleons named in 1843; *ngomeense* is named after the Ngome Forest, where the chameleon was found.
CLASSIFICATION	Animalia • Chordata • Reptilia • Squamata • Chamaeleonidae
TYPE OF MIMICRY	camouflage

SOMSAK PANHA

Shocking Pink Millipede

LET THAT BE A WARNING TO YOU

SOMSAK PANHA

Discovered in 2006, *Desmoxytes purpurosea* is a shocking pink, spiny millipede that is openly active during the day where it can be easily seen—and avoided—by predators. Experienced predators have figured out that bright colors are a signal that potential prey may be harmful or have a foul flavor. Not only does its bright, aposematic coloration warn predators that it's toxic, *D. purpurosea* emits the telltale almondlike odor of cyanide—a chemical that many millipedes use as a chemical defense. Other *Desmoxytes* species from Thailand are strikingly red and are also found in the open. With *D. purpurosea*, these dragon millipedes appear to converge on a common coloration pattern that makes their aposematism even more effective for al fresco dining in the Hup Pa Tard.

DISCOVERED	Hup Pa Tard cavern, Thailand
SCIENTIFIC NAME	*Desmoxytes purpurosea* Enghoff, Sutcharit & Panha 2007
ETYMOLOGY	*Desmoxytes* is a genus of millipedes named in 1923; *purpurosea* is named for the species' purple-pink color, based on a composite of *purpura* and *roseus*, the Latin words for the two colors.
CLASSIFICATION	Animalia • Arthropoda • Diplopoda • Polydesmida • Paradoxosomatidae
TYPE OF MIMICRY	aposematism

THIERRY WEITZMANN

Oria's Leaf Insect

MAKE LIKE A TREE AND LEAVE

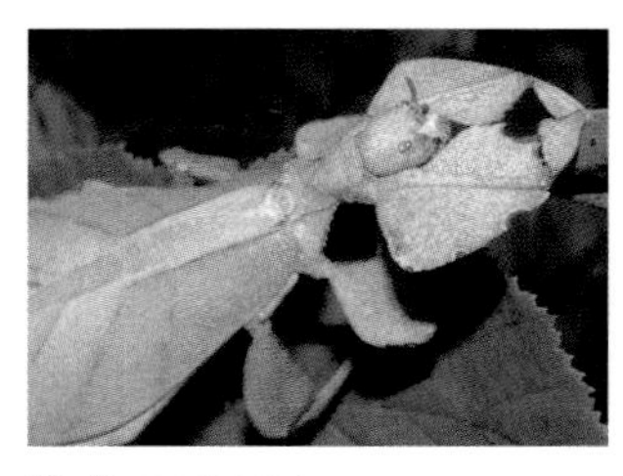

Phyllium ericoriai.

ARNAUD AND CHRISTOPHE BAUDUIN

Now you see it, now you don't. Like other phasmids—stick and leaf insects—Oria's leaf insect shows (or is it hides?) amazing vegetation-like adaptations. The color, texture, shape, and veinlike ridges of the wings, abdomen, and legs combine to give the species a truly leaflike appearance. Although this species had previously been seen, it was long confused with a related species and not recognized as new. Oria's leaf insect has unusually large and complex eggs that are unique among phasmids (which is a feat, considering phasmids are noted for having bizarrely shaped eggs). Newborn nymphs of *Phyllium ericoriai* are also large compared to those of closely related species. This new species was discovered on cultivated guava and mango. Little else is known of the food it eats, and efforts are being made to breed the species in captivity.

Egg of *Phyllium ericoriai.*

FRANK HENNEMANN

DISCOVERED	Marinduque Island, Philippines
SCIENTIFIC NAME	*Phyllium ericoriai* Hennemann, Conle, Gottardo & Bresseel 2009
ETYMOLOGY	*Phyllium* is a genus of leaf insects named in 1798; *ericoriai* is named in honor of Eric Oria, who provided the specimens to the discoverers.
CLASSIFICATION	Animalia • Arthropoda • Insecta • Phasmatodea • Phylliidae
TYPE OF MIMICRY	camouflage

LARRY AND DENISE TACKETT

Denise's Pygmy Seahorse

SEE HORSE?

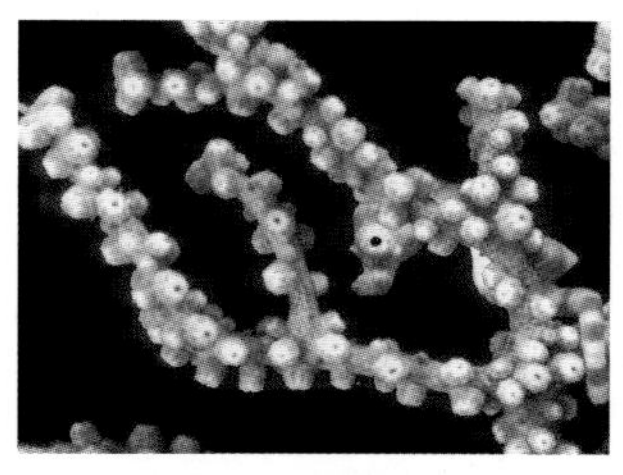

LARRY AND DENISE TACKETT

Many species across the animal kingdom use camouflage to hide from predators, but *Hippocampus denise* is truly a master of disguise. One of the most habitat-savvy animals to be discovered this century, *H. denise* seems able to match any color of the soft coral and gorgonian sea fans among which it lives—including the coral's dots and bumps. With a standard length of 0.52 inches, *H. denise* was considered to be the smallest seahorse in the world when it was described in 2003. Among the smallest vertebrate animals, pygmy seahorses escaped notice due to their size and extraordinary camouflage until 1969. Since that time, nine species have been discovered and named. As marine species exploration continues, it is likely that additional species will be found.

DISCOVERED	Indonesia
SCIENTIFIC NAME	*Hippocampus denise* Lourie & Randall 2003
ETYMOLOGY	*Hippocampus* is a genus of fish named in 1810; *denise* named in honor of Denise Tackett, who brought the species to Lourie's and Randall's attention, who also add that "the name Denise . . . means . . . wild, frenzied," in reference to the activity levels of this new species.
CLASSIFICATION	Animalia • Chordata • Actinopterygii • Syngnathiformes • Syngnathidae
TYPE OF MIMICRY	camouflage

F. SARA CECCARELLI

Jumpin' Spider Ant-Mimic

JUMPING TO CONCLUSIONS

Jumping spiders of the genus *Myrmarachne* are remarkable in their ability to mimic ants in shape and color, and some species can even mimic their social behavior. These spiders simulate the ants that are naturally avoided by the spiders' usual predators and are thus an example of Batesian mimicry. Like other spider relatives in the same genus, new species *M. smaragdina* has an antlike appearance and waves two of its eight legs to look like the antennae of the *Oecophylla smaragdina* ant it mimics (photo at right). Unlike insects with three major body segments, spiders have only two, and these particular spider species create the illusion of a third body segment to better mimic the ant. This new species of jumping spider was collected in 2003.

Ant model *O. smaragdina.*

F. SARA CECCARELLI

DISCOVERED	Australia
SCIENTIFIC NAME	*Myrmarachne smaragdina* Ceccarelli 2010
ETYMOLOGY	*Myrmarachne* is a genus of spiders named in 1838; *smaragdina* is named after its model, the green tree ant *Oecophylla smaragdina*, with the name *smaragdina*, for a word from Arabic meaning emerald, in reference to the green color of the ant's gaster—the bulbous part of the ant's abdomen.
CLASSIFICATION	Animalia • Arthropoda • Arachnida • Araneae • Salticidae
TYPE OF MIMICRY	Batesian

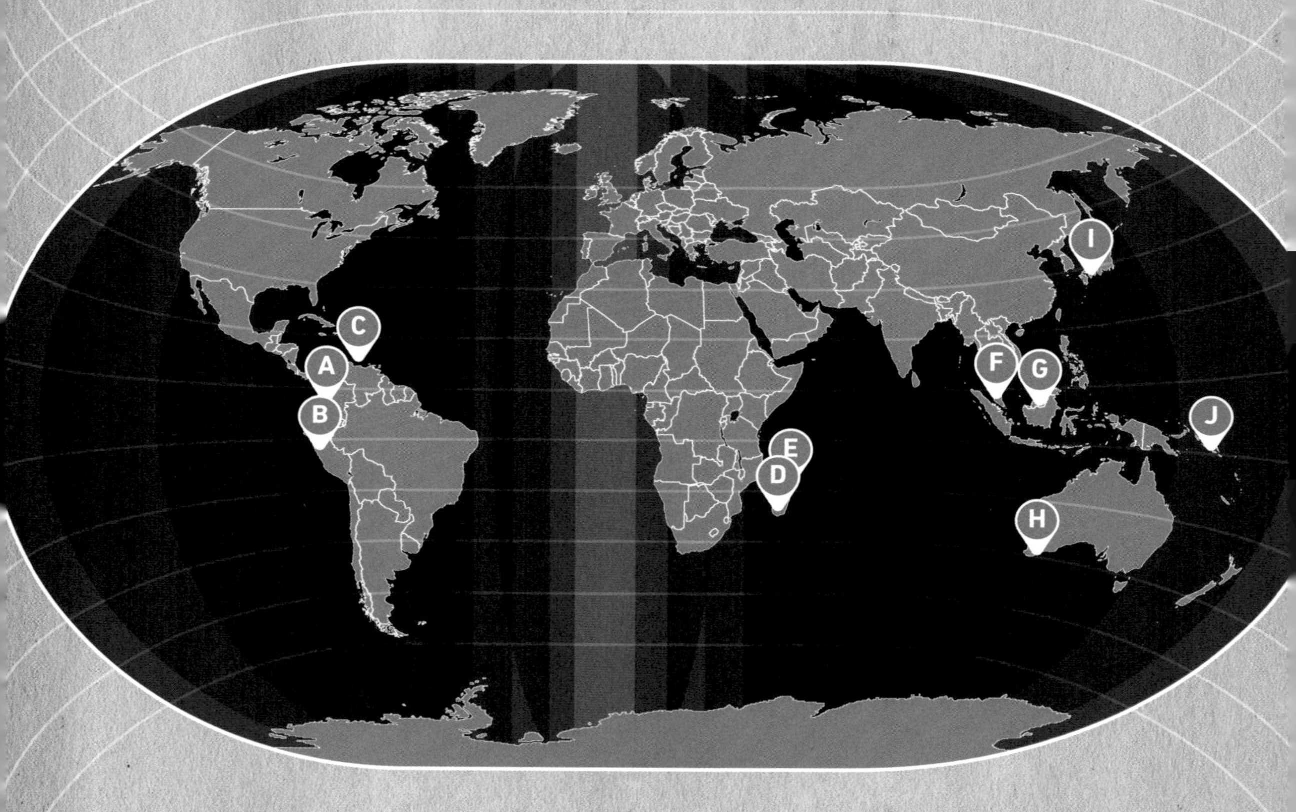

- A Apparating Moon-Gentian
- B Madonna's Water Bear
- C Bonaire Banded Box Jelly
- D John Cleese's Woolly Lemur
- E Google Ant
- F David Bowie's Spider
- G SpongeBob SquarePants Fungus
- H Clare Hannah's Shrimp
- I Groening's Sand Crab
- J Wonderfully Photogenic 'Pus

Chapter 10

WHAT'S IN A NAME?

NEW SPECIES WITH THE BEST NAMES

In this chapter, we present examples of interesting species names, including some that were intended to raise public awareness of the Earth's biodiversity and to help fund species exploration, as well as species that have been named for celebrities, unusual behaviors, bizarre appearances, characters from pop culture, and some of our favorites: names that just sound funny. Species names are known as binominals and consist of two words (both italicized): the genus name (capitalized) plus a *specific epithet* (lowercase) that uniquely identifies each species within that genus. Taxonomists often use descriptive words to help make the name easier to remember and to distinguish one species from another.

The endings of specific epithets sometimes tell us something about the derivation of the names. Under the rules for animal names, a species named after a man typically ends in -*i*, such as *Stenomorpha roosevelti*, a beetle named after President Theodore Roosevelt. A species named after a female ends in -*ae*, such as *Conolophus marthae*, the pink iguana in chapter 6 named after the scientist's daughter. When a species is named to honor a couple, the ending is -*orum*, such as *Orectochilus orbisonorum*, a diving beetle named after Roy and Barbara Orbison. The suffix -*ensis* tells us that a species is named

COURTESY OF VINTAGE PRINTABLE

for a place, usually the place where the species was first found. For example, *Pterourus appalachiensis*, the butterfly in chapter 9 that was found in the Appalachian Mountain region of the southeastern United States.

Giving species weird or funny names isn't as rare as you might think. Smithsonian scientist Terry Erwin is hard to match for groan-inducing puns from his studies of the beetle genus *Agra* in which he named species *Agra vation*, *Agra cadabra*, and *Agra phobia*. Entomologist Gordon Marsh named two wasps *Heerz lukenatcha* and *Heerz tooya* ("Here's lookin' at ya" and "Here's to ya") and malacologist Alan Solem must have been thinking of *A Christmas Carol* when he named a snail *Ba humbugi*. For those into the Roman classics, Paul Spangler named a beetle *Ytu brutus* and David Steadman and Marie Zarriello named an extinct parrot *Vini vidivici* as a modern wordplay on "I came, I saw, I conquered." Not to be outdone, fly expert Neal Evenhuis gave us *Pieza kake*, *Pieza pi*, *Pieza rhea* (piece of cake, pizza pie, pizzeria), and, of course, *Pieza deresistans*. A discussion of humorous names would also not be complete without bathroom humor, such as *Colon rectum* (a beetle) and *Aploparaksis turdi* (a tapeworm that is found, where else, in feces). And, a tiny mushroom in the *Phallus* genus, shaped like a penis, was named after a colleague with his full knowledge and permission—nothing diminutive about his sense of humor!

While most of the above examples are from recent decades, taxonomists have had a sense of humor from the beginning. George Willis Kirkaldy was criticized by the London Zoological Society in 1912 for his frivolity in naming bugs *Polychisme*, *Ochisme*, and *Peggichisme*, which must have brought a blush to Victorian cheeks with all of that smooching. Hopefully Polly and Peggy never knew about each other.

Other names can have an edge to them or are downright insulting. One of the most notorious scientific feuds was between dinosaur hunters E. D. Cope and O. C. Marsh.

Marsh named *Mosasaurus copeanus* in "honor" of Cope. Cope named *Anisonchus cophater*, a fossil mammal, in recognition of the Cope haters surrounding him. Later, L. M. van Valen named another mammal *Oxyacodon marshater* just to even the score. This tradition of attacking one's enemies with species names began with Linnaeus himself in the eighteenth century. Linnaeus named the flower genus *Commelina* after three members of the Dutch Commelin family, of which only two were successful. The flower? It has three petals, one of which is small, pale, and shriveled. In another case, Linnaeus's student Daniel Rolander collected specimens for Linnaeus in Suriname but then refused to hand them over. In response, Linnaeus took the Greek word *aphanus*, meaning ignoble or inferior, and coined *Aphanus rolandri*. Take that.

Carolus Linnaeus by Per Krafft the Elder (1774).

JASON R. GRANT

Apparating Moon-Gentian

HARRY POTTER ALLUSION ILLUSION

With a leaf that looks like tobacco and a flower too big to fit on a folded newspaper (eleven by seventeen inches), this new plant species is a sturdy, flowering tree that can grow over sixteen feet high. Not only are its flowers large, but the tree seems to magically bloom just after twilight and suddenly materialize out of nowhere in the darkening woods. The scientists who found this new species decided to name it after the wizardly mode of transportation called apparation that "appeared" in the Harry Potter books and movies. This new moon-gentian species was discovered—or apparated—in the neotropical forests of Podocarpus National Park, and it's one of thirty or more species of the biodiverse moon-gentians that are found in southern Ecuador. Dumbledore would most definitely be pleased.

JASON R. GRANT

DISCOVERED	Podocarpus National Park, Ecuador
SCIENTIFIC NAME	*Macrocarpaea apparata* Grant & Struwe 2003
ETYMOLOGY	*Macrocarpaea* is a genus of flowering plants named in 1895; *apparata* is named after the English word "apparate," meaning to magically appear.
CLASSIFICATION	Plantae • Eudicots • Asterids • Gentianales • Gentianaceae

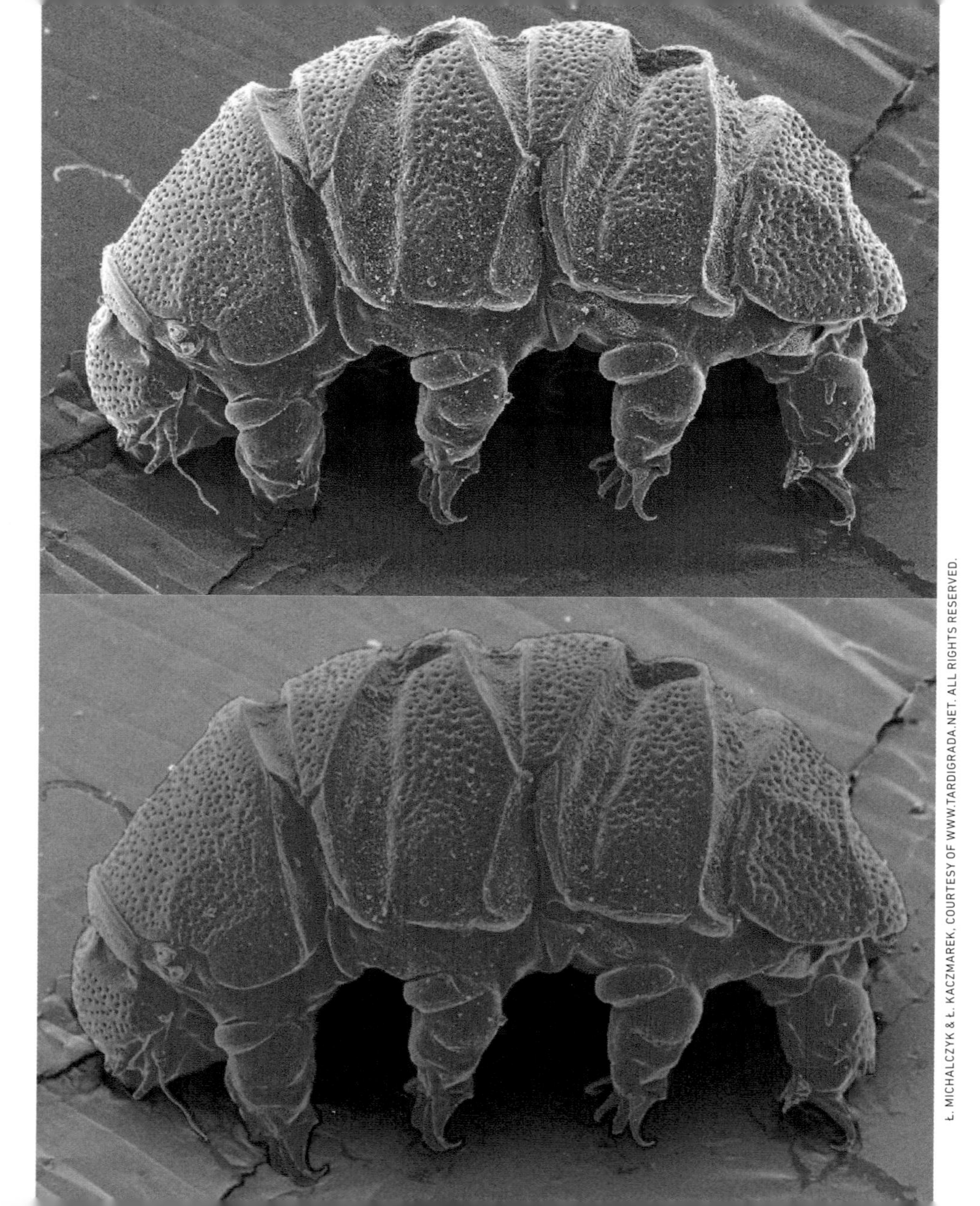

Madonna's Water Bear

PIGS IN SPACE!

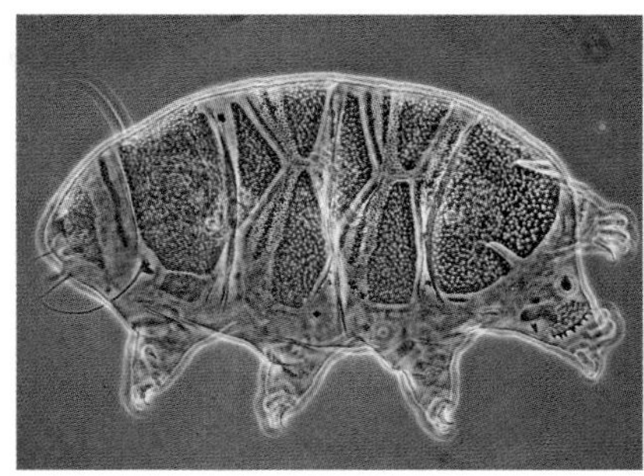

Ł. MICHALCZYK & Ł. KACZMAREK, COURTESY OF WWW.TARDIGRADA.NET.

Naming a new species after someone famous not only allows a scientist to honor his or her favorite celebrity but it also helps bring attention to Earth's biodiversity by increasing awareness about little-known species that can be as beautiful and engaging as their namesakes. This new tardigrade, *Echiniscus madonnae*, was named after Madonna, "one of the most significant artists of our times," according to the scientists who discovered it. Typically measuring about 0.04 inches, tardigrades, also called moss piglets and water bears, are microscopic and practically indestructible animals. They are eight-legged creatures with tiny claws and they lumber like a bear as they walk. They live in water sediments, lichens, and moss and can be found in some of Earth's most extreme environments, from the polar caps to high elevations. In fact, tardigrades are so durable that they've been launched into orbit around Earth, where they have survived the radiation and vacuum of space. Thus, Madonna's water bear is a wonderful example of a species that is out of this world.

DISCOVERED	Peru
SCIENTIFIC NAME	*Echiniscus madonnae* Michalczyk & Kaczmarek 2006
ETYMOLOGY	*Echiniscus* is a genus of tardigrades named in 1840; *madonnae* is named in honor of singer Madonna.
CLASSIFICATION	Animalia • Tardigrada • Heterotardigrada • Echiniscoidea • Echiniscidae

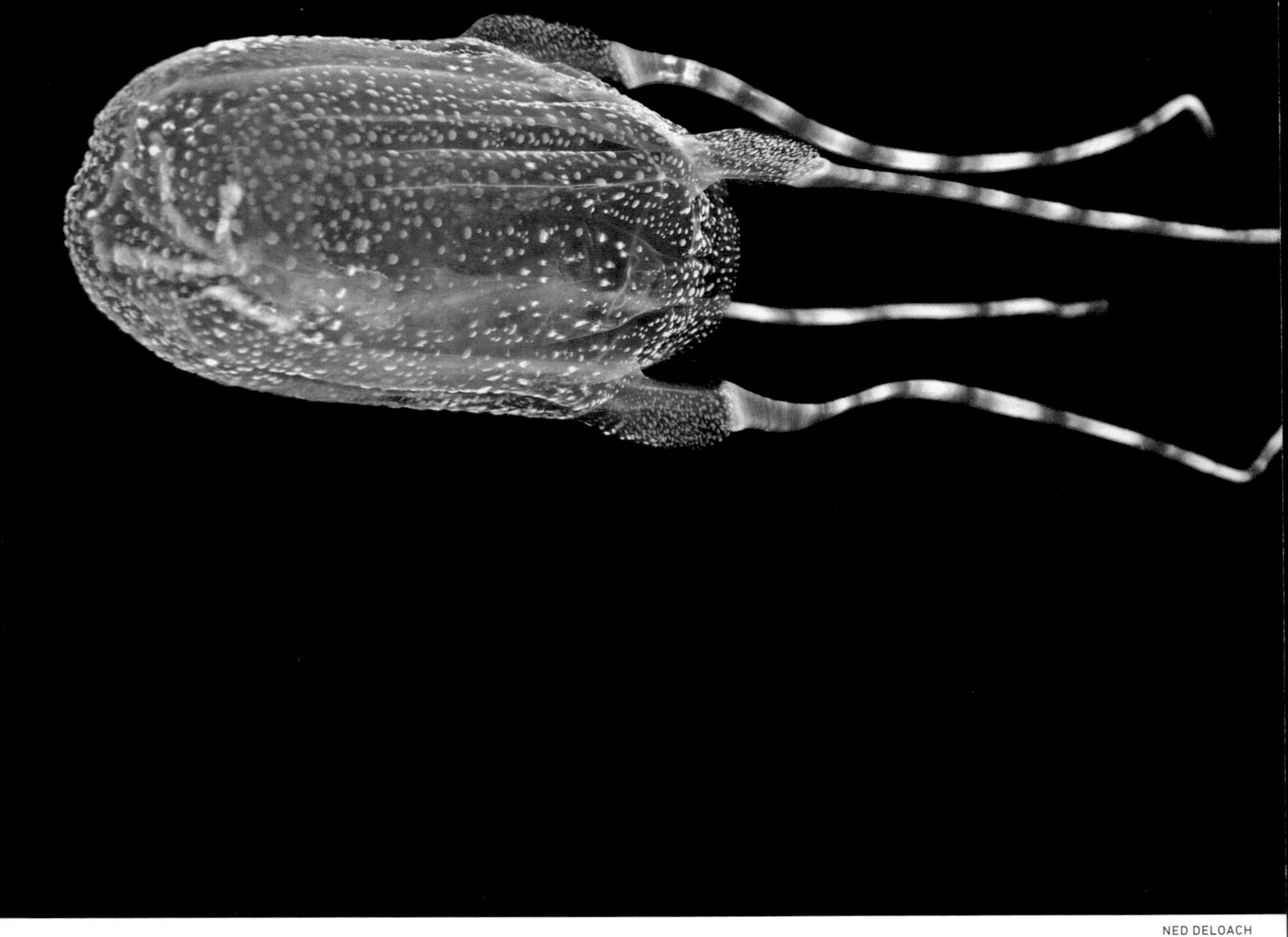

NED DELOACH

Bonaire Banded Box Jelly

OH BOY!

This strikingly beautiful and unusually banded box jelly had so many sightings during the past decade that it had been given a common name long before it was officially described in 2011, three years after a specimen was finally captured and provided to scientists. The numerous sightings of this new species illustrate the importance of citizen science and how the public can be involved in biodiversity exploration and in helping to name new species. More than three hundred entries were submitted in an online competition to name this *Tamoya* jelly, and hundreds of votes were cast to select *ohboya*. The winning name was suggested by Lisa Peck, a high school biology teacher. Ms. Peck based her entry on how people must exclaim "Oh boy!" when first encountering this amazing jelly—including swimmers, scuba divers, scientists, and even the doctors who have treated victims of *Tamoya ohboya*'s venomous stings.

MARIKE WILHELMUS

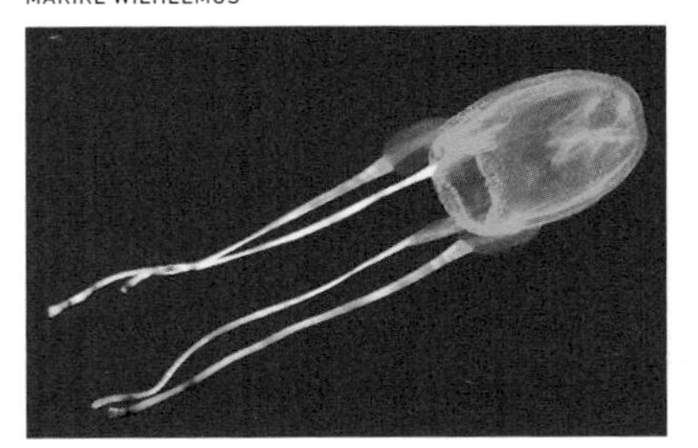

JIM PLATZ

DISCOVERED	Bonaire, Netherlands (Dutch Caribbean)
SCIENTIFIC NAME	*Tamoya ohboya* Collins, Gillan, Lynn, Morandini & Marques 2011
ETYMOLOGY	*Tamoya* is a genus of box jellyfish named in 1859; *ohboya* in reference to "Oh boy!" for the likely reaction of someone on first encounter with this species.
CLASSIFICATION	Animalia • Cnidaria • Cubozoa • Carybdeida • Tamoyidae

URS THALMANN

John Cleese's Woolly Lemur

BORN TO BE WILD

Discovered in 2005, this new woolly lemur is a fitting tribute to John Cleese, self-proclaimed British "writer, actor, and tall person," who made cute furry animals a cultural icon in *Monty Python and the Holy Grail*. Who could forget the white bunny that guarded the Holy Grail with "nasty, big, pointy teeth" and a "vicious streak a mile wide"? Poking fun at cute furry mammals with killer instincts may be his day job, but John Cleese is also a serious species advocate who has produced such conservation films as *Fierce Creatures*, a comedy, and *In the Wild: Lemurs with John Cleese*, a documentary. John Cleese's woolly lemur was thought to be so critically endangered that the scientists who named this species considered it unethical to capture an animal and allow it to die in captivity for additional scientific study. Instead, the scientists compared the new species to known woolly lemurs and based the description and naming of this new species on collected hair samples, photographs, videotapes, and anesthetized animals.

URS THALMANN

DISCOVERED	Bemaraha, Madagascar
SCIENTIFIC NAME	*Avahi cleesei* Thalmann & Geissmann 2005
ETYMOLOGY	*Avahi* is a genus of woolly lemurs named in 1834; *cleesei* is named after John Cleese to honor his contributions to the promotion of conservation issues.
CLASSIFICATION	Animalia • Chordata • Mammalia • Primates • Indriidae

ANTWEB.ORG

Google Ant

SIX-LEGGED SEARCH ENGINES

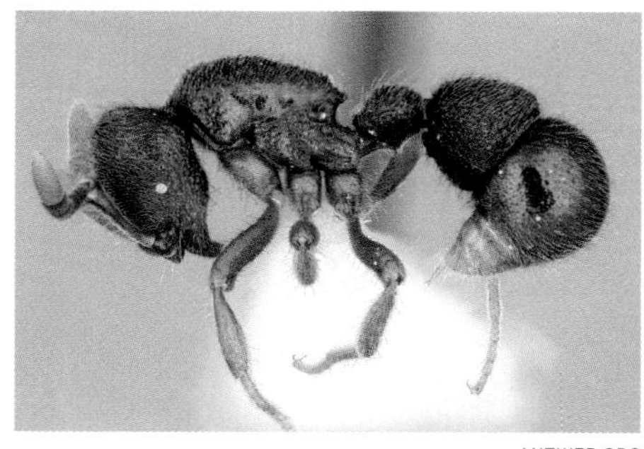
ANTWEB.ORG

Dr. Brian Fisher of the California Academy of Sciences, San Francisco, named this species of ant in honor of the Internet search engine Google, citing the importance of Google Earth software in helping to connect people worldwide to biodiversity. The satellite images of this software allow scientists to scan Earth's landscape for potential field sites and promising diversity hot spots, while helping people visualize the geographical and ecological distribution of any species by simply locating the GPS coordinates where a species is known to live. The diversity of ant species can be explored by looking up a name on Antbase.org or downloading a list of all the species found at a location—whether it's in your backyard or some remote tropical forest. Workers of the new ant species are themselves tireless little search engines, constantly tracking down the unusual delicacy that forms their diet of spider eggs.

DISCOVERED	Madagascar
SCIENTIFIC NAME	*Proceratium google* Fisher 2005
ETYMOLOGY	*Proceratium* is a genus of ants named in 1863; *google* is named in recognition of the Internet search engine and mapping technologies of Google and to represent the ant's specialized ability to search for spider eggs that are difficult to find.
CLASSIFICATION	Animalia • Arthropoda • Insecta • Hymenoptera • Formicidae

PETER JÄGER, SENCKENBERG FRANKFURT

David Bowie's Spider

SPIDERS FROM ~~MARS~~ MALAYSIA

Heteropoda davidbowie is a new species of medium- to large-sized spider that was found between 2004 and 2008 in several parts of Southeast Asia, including Malaysia, Thailand, Indonesia, and Singapore. Named in honor of rock-and-roller David Bowie, this new spider is strikingly beautiful, with bright orange hair covering its body and shiny black and yellow jaws. Females are larger than males and may reach one inch in length. David Bowie's album *The Rise and Fall of Ziggy Stardust and the Spiders from Mars* and his song "Glass Spider" were part of the inspiration for the name. In the same journal article where David Bowie's spider was published, twenty-four other new spider species were described, including another long-haired species named *H. hippie*—to which we say, "Far out."

PETER JÄGER, SENCKENBERG FRANKFURT

DISCOVERED	Malaysia
SCIENTIFIC NAME	*Heteropoda davidbowie* Jäger 2008
ETYMOLOGY	*Heteropoda* is a genus of spiders named in 1804; *davidbowie* is named after singer David Bowie.
CLASSIFICATION	Animalia • Arthropoda • Arachnida • Araneae • Sparassidae

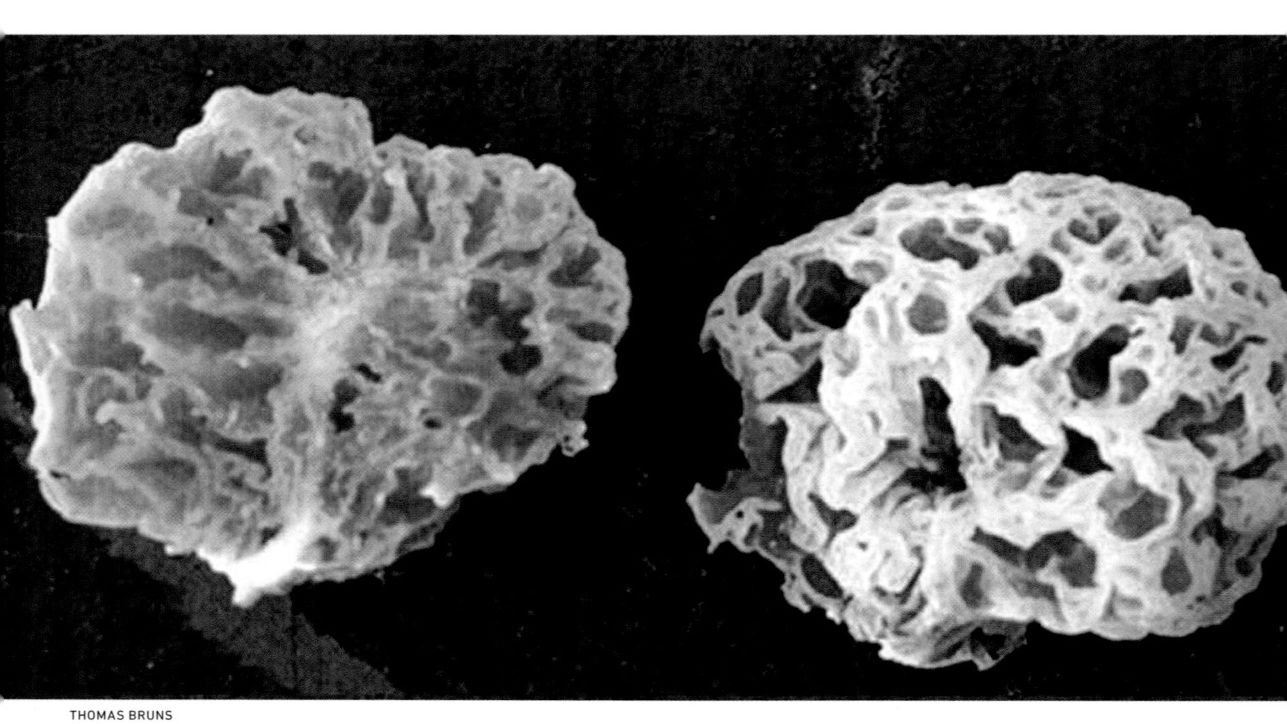
THOMAS BRUNS

SpongeBob SquarePants Fungus

BIKINI BOTTOM BOLETE

DENNIS E. DESJARDIN & ANDREW ICHIMURA

Discovered in 2010, this new species of bolete fungus was found growing on the ground in the tropical rain forest of Borneo's Lambir Hills National Park. Bolete fungi are mushrooms that have a spongelike surface under the cap rather than gills. According to the scientists who discovered this new species, *Spongiforma squarepantsii* is unique for a mushroom since it "can be squeezed to wring out water and it springs back to its normal shape and size" like a sponge. It was named after the cartoon character SpongeBob SquarePants not only because of its spongelike qualities but also because it smells fruity and, as we know, SpongeBob lives in a pineapple. When the mushroom is viewed with a scanning electron microscope, its surface looks like a "seafloor covered with tube sponges," which also resembles SpongeBob's home in Bikini Bottom. Although the species' name was originally rejected by journal editors as "frivolous," the scientists thankfully persisted, to which we say, "Holy Krabby patties!" for bringing attention to the biodiversity of forest species around the world.

DISCOVERED	Lambir Hills National Park, Borneo
SCIENTIFIC NAME	*Spongiforma squarepantsii* Desjardin, Peay & T. D. Bruns 2011
ETYMOLOGY	*Spongiforma* is a genus of spongelike fungi named in 2009 by Desjardin, Binder, Roekring & Flegel; *squarepantsii* is named for a resemblance to the well-known cartoon character SpongeBob SquarePants.
CLASSIFICATION	Fungi • Basidiomycota • Agaricomycetes • Boletales • Boletaceae

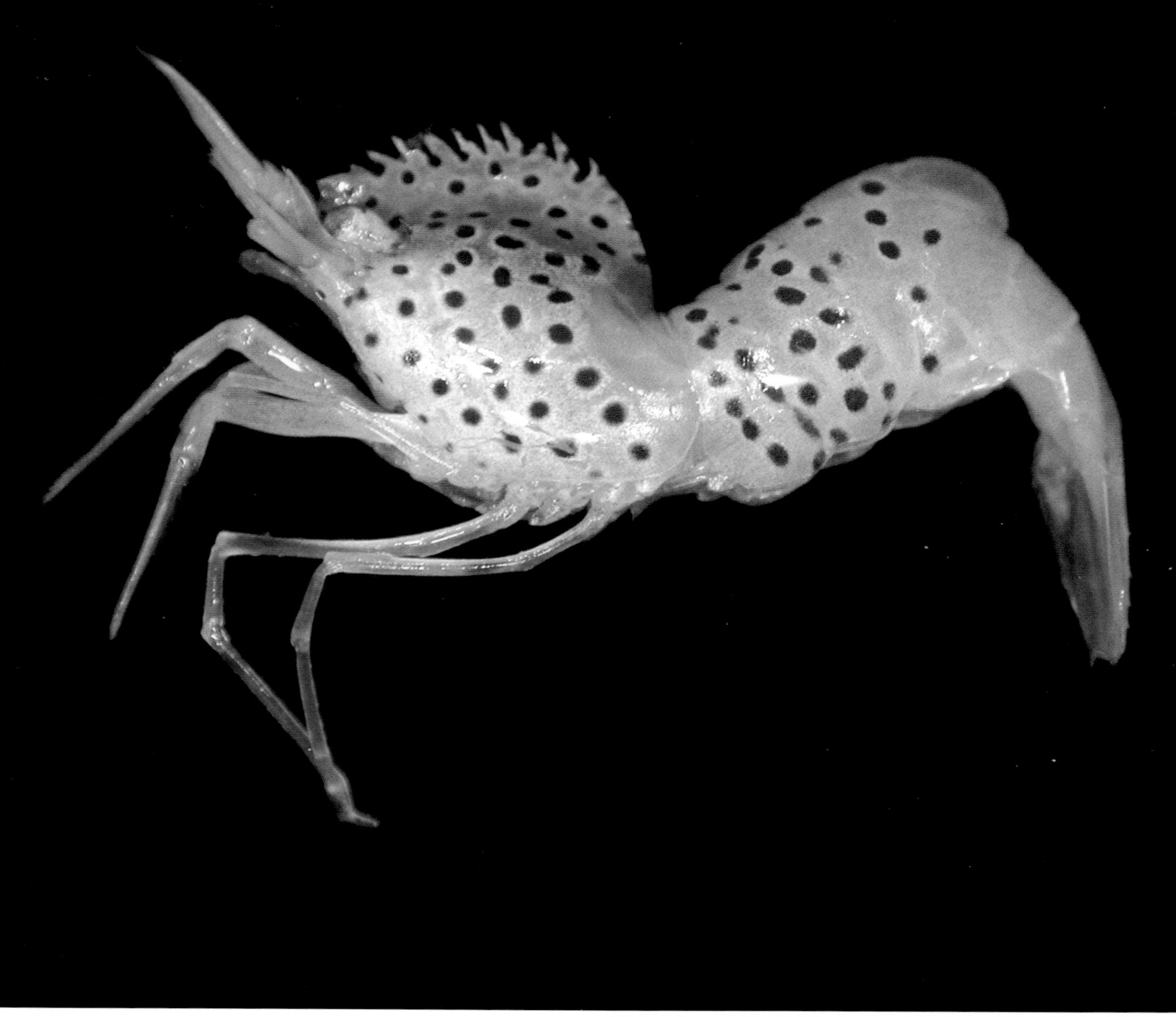

Clare Hannah's Shrimp

eBAY eTYMOLOGY

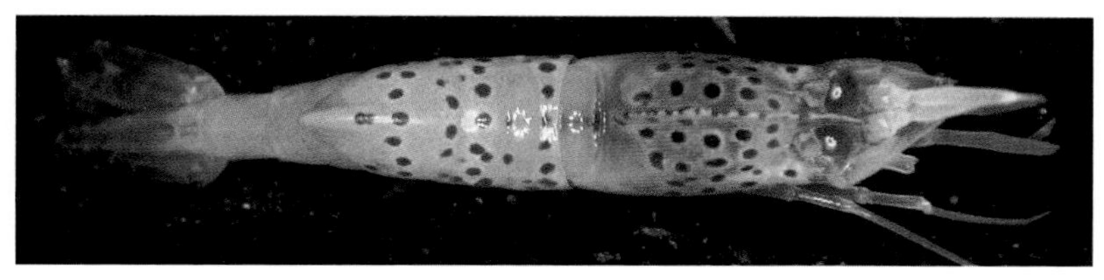

© KAREN GOWLETT-HOLMES, CSIRO

Covered in red spots and described as "jewel-like" with a jagged crest that looks like a Mohawk, this beautiful and unique-looking shrimp was christened by an NBA star who won the naming rights by having the highest bid on eBay. *Lebbeus clarehannah* was discovered in 2005 by Anna McCallum, who championed the idea of having an eBay charity auction as a way to raise money for marine conservation and to increase public awareness of Australia's ocean species. The proceeds from the auction helped to protect Australia's southwest coast, an area where the vast majority of its marine species live nowhere else on Earth. In 2009, the right to name this new shrimp was "sold!" to Australian Luc Longley, formerly of the NBA's Chicago Bulls, who named it after his daughter Clare Hannah. Longley definitely scored one for biodiversity *and* the home team.

DISCOVERED Western Australia

SCIENTIFIC NAME *Lebbeus clarehannah* McCallum & Poore 2010

ETYMOLOGY *Lebbeus* is a genus of shrimp named in 1847; *clarehannah* after Clare Hannah Longley.

CLASSIFICATION Animalia • Arthropoda • Crustacea • Malacostraca • Decapoda • Hippolytidae

TIN-YAM CHAN, NATIONAL TAIWAN OCEAN UNIVERSITY

Groening's Sand Crab

AY CARUMBA!

Just to give you a historical perspective on how long it can take to make a new species known, specimens of this unusual crab have been stored in museum collections for decades even though the crab was not described and named until 2002—seventy years after the first specimen was collected and deposited at Harvard University's Museum of Comparative Zoology. The "holotype"—the specimen that bears the name of the species—was collected in waters off Japan about three months before the attack on Pearl Harbor in 1941 and later stored at the National Museum of Natural History in the Netherlands. The species is named after Matt Groening, cartoonist and creator of *The Simpsons*. The species' discoverers noted that Groening and *The Simpsons* have "exposed people to a diversity of crustacean species" including Lisa's hermit crab and "Pinchy" the lobster. Given this crab's history of sitting around on a shelf, it's no wonder Bart Simpson said, "I smell a museum."

PKL NG, RAFFLES MUSEUM, NUS

Beach habitat of *Albunea groeningi*, Western Australia.

A. HOSIE, WESTERN AUSTRALIAN MUSEUM

DISCOVERED	Philippines and Japan
SCIENTIFIC NAME	*Albunea groeningi* Boyko 2002
ETYMOLOGY	*Albunea* is a genus of crabs named in 1795; *groeningi* is named in honor of Matt Groening, American screenwriter, producer, and animation artist.
CLASSIFICATION	Animalia • Arthropoda • Crustacea • Malacostraca • Decapoda • Albuneidae

ERIC HOCHBERG

Wonderfully Photogenic 'Pus

PAPARAZZI'D OCTOPUS

MARK NORMAN

If there were octopus celebrity magazines, *Wunderpus photogenicus* would be on every cover. Even before it was recognized as a new species, photographs of this gorgeous octopus were being taken by divers who marveled at its incredible beauty and had given it a common name, "Wonderpus." *W. photogenicus* has the incredible ability to detach one of its arms (or more as needed) and trick a predator into attacking the severed limb as a wiggly decoy—similar to a submarine releasing countermeasures to misdirect a torpedo's sonar. Beautiful, talented, *and* photogenic—no wonder this new species is so wonderfully named.

DISCOVERED	Indonesia
SCIENTIFIC NAME	*Wunderpus photogenicus* Hochberg, Norman & Finn 2006
ETYMOLOGY	*Wunderpus* after its common name of "Wonderpus" and based on the German word *wünder*, meaning marvel or wonder, plus the Greek *pous*, meaning foot; *photogenicus* relates to the species' "photogenic" qualities and reflects the immense photographic interest in this octopus.
CLASSIFICATION	Animalia • Mollusca • Cephalopoda • Octopoda • Octopodidae

Conclusion

Each species that a scientist discovers adds something new and unexpected to our knowledge of Earth while allowing us to understand and appreciate the diversity of life around us just a little bit more. Among the new species found every year are the beautiful and ugly, the benign and threatening, and the flourishing and threatened. It is through the discovery, description, and naming of species that we ultimately understand the origin and history of life on this remarkable planet we call home.

The way in which we respect and treat the other species on this planet and respond to the biodiversity crisis says much about our humanity. Although the increasing support for the ideals of species conservation and sustainable biodiversity are encouraging, we cannot even begin to value or save species if we do not know where they live on the globe or even if they exist at all. It is through species exploration that we empower people to experience, enjoy, and be inspired to preserve Earth's biodiversity. As humans, we have a choice with enormous consequences: we can either accelerate species exploration to measure our progress toward sustainable biodiversity or we can simply take our chances and hope for the best as millions of species are threatened by possible extinction.

When Linnaeus set out to catalog species in 1758, his vision was way ahead of its time. Until the latter half of the twentieth century, scientists did not have the travel, communication, or data management capabilities to deal with the millions of species distributed across the face of our planet. Technological advances—particularly in the computer and information sciences—give us the means today to mount an aggressive scientific mission to discover and describe all species and to perhaps finish the inventory that Linnaeus began.

Our hope is that the twenty-first century will be known as the golden age of species discovery. At this time, science has documented only about 2 million species and at least 10 million others are waiting to be discovered. Many species that can help us in our conservation goals and to understand our origins are about to disappear. We only need to act on our curiosity. It is our hope that the species in this book will intrigue and inspire you to explore your local natural history museum or botanical garden or aquarium. Perhaps you'll add some new plants to your own garden that will attract insect or mollusk species. Or maybe you'll be inspired to buy a field guide and take a hike on a nature trail.

It is as simple as learning to cross a street: Stop. Look. Listen. You just never know what amazing species you may find.

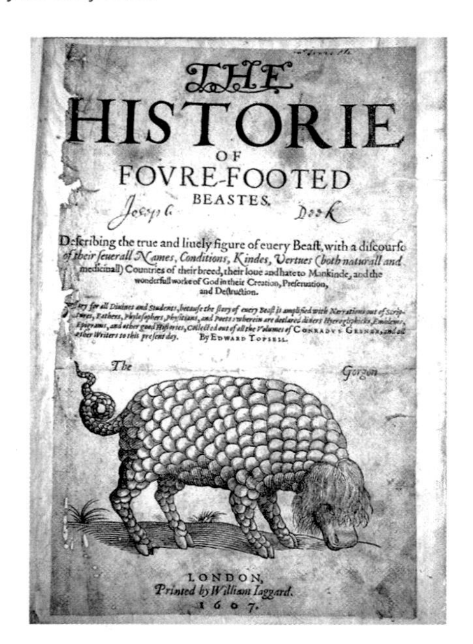

THE
HISTORIE
OF
FOVRE-FOOTED
BEASTES.

Describing the true and liuely figure of euery Beast, with a discourse of their seuerall Names, Conditions, Kindes, Vertues (both naturall and medicinall) Countries of their breed, their loue and hate to Mankinde, and the wonderfull worke of God in their Creation, Preseruation, and Destruction.

...ry for all Diuines and Students, because the story of euery Beast is amplified with Narrations out of Scriptures, Fathers, Phylosophers, Physitians, and Poets: wherein are declared diuers Hyeroglyphicks, Emblems, Epigrams, and other good Histories, Collected out of all the Volumes of CONRADVS GESNER, and all other Writers to this present day. By EDWARD TOPSELL.

The Gorgon

LONDON,
Printed by William Iaggard.
1607.

Acknowledgments

We wish to first thank all of the scientists and photographers who so willingly and graciously contributed their images, time, and efforts that allowed us to feature the amazing species that appear in this book. Without their generous help and wonderful discoveries, this project would not have been possible.

Our deepest gratitude goes to the Plume imprint of the Penguin Group for providing the opportunity to share our passion and our stories about Earth's species and biodiversity. We are incredibly grateful to our editors Nadia Kashper and Kate Napolitano for helping us pull and push the sled uphill so we can have that thrilling ride back down. An especially big thanks goes to Kate for her well-grounded edits, indefatigable encouragement, unfailing answers, and for guiding this project to completion.

Many thanks also to Rebecca Dornburg and Michael Dambrowski for their graphics help; to Christopher Miller, geographer extraordinaire, for mapping species locations; and to Nico Franz for his support and generous availability on last-minute consultations. Finally, we thank our families and friends who listened to all of our stories and shared our excitement about the species in this book. Your encouragement and enthusiasm kept us going and on course.

Thank you, all.

References and Credits

CHAPTER 1. PRETTY COUNTS: THE PRETTIEST NEW SPECIES

Kaiser's Nudibranch

Hermosillo, A. and Á. Valdés (2007). "A new *Polycera* (Opisthobranchia: Mollusca) from Bahía de Banderas, México." *Proceedings of the California Academy of Sciences* 58: 477–484.

Museums of Record: California Academy of Sciences, San Francisco, California, USA; Natural History Museum of Los Angeles County, Los Angeles, California, USA

Paisa Orchid Bee

Ramírez, S. (2005). "*Euglossa paisa*, a new species of orchid bee from the Colombian Andes (Hymenoptera: Apidae)." *Zootaxa* 1065: 51–60.

Museum of Record: Collection Alexander von Humboldt, Bogotá, D.C., Colombia

Kovach's Orchid

Atwood, J. T., S. Dalström, and R. Fernandez (2002). "*Phragmipedium kovachii*, a new species from Peru." *Selbyana* 23, Supplement: 1–4.

Institution of Record: United States Botanic Garden, Washington, D.C., USA

Diamantina Tarantula

Bertani, R., T. dos Santos, and A. F. Righi (2009). "A new species of *Oligoxystre* Vellard 1924 (Araneae, Theraphosidae) from Brazil." *ZooKeys* 5: 41–51.

Museum of Record: Instituto Butantan and the Museu de Zoologia da Universidade de São Paulo, São Paulo, Brazil.

Patton's Bright Snake

Vieites, D. R., F. M. Ratsoavina, R.-D. Randrianiaina, Z. T. Nagy, R. Glaw, and M. Vences (2010). "A rhapsody of colours from Madagascar: discovery of a remarkable new snake of the genus *Liophidium* and its phylogenetic relationships." *Salamandra* 46(1): 1–10.

Museum of Record: Zoologische Staatssammlung Munchen, Münich, Germany

Fried Eggs Worm

James, S. W. (2009). "Revision of the earthworm genus *Archipheretima* Michaelsen (Clitellata: Megascolecidae), with descriptions of new species from Luzon and Catanduanes Islands, Philippines." *Organisms Diversity & Evolution* 9: 244.e1–244.e16.

Museums of Record: Annelida Collection, National Museum of the Philippines, Manila, Philippines; University of Kansas Natural History Museum, Lawrence, Kansas, USA

Psychedelic Frogfish

Pietsch, T. W., R. J. Arnold, and D. J. Hall (2009). "A bizarre new species of frogfish of the genus *Histiophryne* (Lophiiformes: Antennariidae) from Ambon and Bali, Indonesia." *Copeia* 1: 37–45.

Museums of Record: Pusat Penelitian dan Pengembangan Oseanologi, Jakarta, Indonesia; the Burke Museum of Natural History and Culture, University of Washington, Seattle, Washington, USA

Wilson's Blue-Eyed Cuscus

Helgen, K. M., and T. F. Flannery (2004). "Notes on the phalangerid marsupial Genus *Spilocuscus*, with description of a new species from Papua." *Journal of Mammalogy* 85(5): 825–833.

Museum of Record: Nationaal Natuurhistorisch Museum (formerly Rijksmuseum van Natuurlijke Historie), Leiden, the Netherlands

Barbie Pagoda Fungus

Ducousso, M., S. Proust, D. Vigier, and G. Eyssartier (2009). "*Podoserpula miranda* prov. name, a spectacular new fungus species discovered in New-Caledonia (*Podoserpula miranda* nom prov. une nouvelle espèce de champignon très spectaculaire découverte en Nouvelle-Calédonie)." *Bois et forêts des tropiques* 302(4): 73–75.

Museums of Record: Institute for Research and Development *Laboratoire des Symbioses Tropicales* in collaboration with the Muséum National d'Histoire Naturelle, Paris, France

Exquisite Sea Urchin

Coppard, S. E., and H. A. G. Schultz (2006). "A new species of *Coelopleurus* (Echinodermata: Echinoidea: Arbaciidae) from New Caledonia." *Zootaxa* 1281: 1–19.

Museum of Record: The Muséum National d'Histoire Naturelle, Paris, France

CHAPTER 2. STRANGER THAN (SCIENCE) FICTION: THE STRANGEST NEW SPECIES

Little Fork Orchid

Dressler, R. L., and D. Bogarin (2007). "A new and bizarre species in the genus *Condylago* (Orchidaceae: Pleurothallidinae) from Panama." *Harvard Papers in Botany* 12(1): 1–5.

Institution of Record: University of Panama, Panama City, Panama

Quechuan Broad-Nosed Bat

Velazco, P. M. (2005). "Morphological phylogeny of the bat genus *Platyrrhinus* Saussure, 1860 (Chiroptera: Phyllostomidae) with the description of four new species." *Fieldiana: Zoology, New Series* 105: 1–54.

Museums of Record: Field Museum of Natural History, Chicago, Illinois, USA; National Museum of Natural History, Smithsonian Institution, Washington, D.C., USA

Gomes's Freshwater Stingray

De Carvalho, M. R., and N. R. Lovejoy (2011). "Morphology and phylogenetic relationships of a remarkable new genus and two new species of neotropical freshwater stingrays from the Amazon basin (Chondrichthyes: Potamotrygonidae)." *Zootaxa* 2776: 13–48.

Museums of Record: Museu de Zoologia da Universidade de São Paulo, São Paulo, Brazil; American Museum of Natural History, New York, New York, USA

Double-Hooked Anglerfish

Pietsch, T. W. (2005). "New species of the ceratioid anglerfish genus *Lasiognathus* Regan (Lophiiformes: Thaumatichthyidae) from the eastern North Atlantic off Madeira." *Copeia*: 77–81.

Museum of Record: Natural History Museum, London, England, UK

Dumbo Octopus

Collins, M. A. (2003). "The genus *Grimpoteuthis* (Octopoda: Grimpoteuthidae) in the north-east Atlantic, with descriptions of three new species." *Zoological Journal of the Linnean Society* 139: 93–127.

Museum of Record: Natural History Museum, London, England, UK

Big Brain Protist

Gooday, A. J., A. A. da Silva, and J. Pawlowski (2011). "Xenophyophores (Rhizaria, Foraminifera) from the Nazaré Canyon (Portuguese margin, NE Atlantic)." *Deep-Sea Research II* 58: 2401–2419.

Museum of Record: Natural History Museum, London, England, UK

Little Grooves Earthstar

Zamora, J. C., F. D. Calonge (2007). "*Geastrum parvistriatum*, a new species found in Spain (*Geastrum parvistriatum*, una nueva especie encontrada en España)." *El Boletín de la Sociedad Micològica de Madrid* 31: 139–149.

Museums of Record: Real Jardín Botánico, Madrid, Spain; Royal Botanic Gardens, Kew, London, UK; Laboratoire de Cryptogamie, Muséum National d'Histoire Naturelle, Paris, France

Long-Neck Assassin Spider

Wood, H. (2008). "A revision of the assassin spiders of the *Eriauchenius gracilicollis* group, a clade of spiders endemic to Madagascar (Araneae: Archaeidae)." *Zoological Journal of the Linnean Society* 152: 255–296.

Museum of Record: California Academy of Sciences, San Francisco, California, USA

Sahyadri Nose Frog

Biju, S. D., and F. Bossuyt (2003). "New frog family from India reveals an ancient biogeographical link with the Seychelles." *Nature* 425: 711–714.

Museum of Record: Bombay Natural History Society, Bombay, India

Ausubel's Mighty Claw Lobster

Ahyong, S. T., T.-Y. Chan, and P. Bouchet (2010). "Mighty claws: a new genus and species of lobster from the Philippine deep sea (Crustacea, Decapoda, Nephropidae)." *Zoosystema* 32(3): 525–535.

Museums of Record: Crustacean Collection National Museum of the Philippines, Manila, Philippines; Australian Museum, Sydney, Australia

CHAPTER 3. LESS IS MORE, MORE OR LESS: THE SMALLEST NEW SPECIES

Roosmalen's Hairy Dwarf Porcupine

Voss, R. S., and M. N. F. da Silva (2001). "Revisionary notes on neotropical porcupines (Rodentia: Erethizontidae). 2. A review of the *Coendou vestitus* group with descriptions of two new species from Amazonia." *American Museum Novitates* No. 3351: 1–36.

Institute of Record: Instituto Nacional de Pesquisas da Amazonia, Manaus, Brazil

Vanessa's Bamboo

Judziewicz, E. J., and S. Sepsenwol (2007). "The world's smallest bamboo: *Raddiella vanessiae* (Poaceae: Bambusoideae: Olyreae:), a new species from French Guiana." *Journal of the Botanical Research Institute of Texas* 1(1): 1–7.

Museums of Record: National Museum of Natural History, Smithsonian Institution, Washington, D.C., USA; Herbier de Guyane, Institut de recherche pour le développement, Cayenne, French Guiana

Pernambuco Pygmy-Owl

da Silva, J. M. C., G. Coelho, and L. P. Gonzaga (2002). "Discovered on the brink of extinction: a new species of Pygmy-Owl (Strigidae: *Glaucidium*) from Atlantic Forest of northeastern Brazil." *Ararajuba* 10(2): 123–130.

Institute of Record: Ornithological Collection, Universidade Federal de Pernambuco, Recife, Brazil

Smallest Crustacean

Knudsen, S. W., M. Kirkegaard, and J. Olesen (2009). "The tantulocarid genus *Arcticotantalus* removed from Basipodellidae into Deoterthridae (Crustacea: Maxillopoda) after the description of a new species from Greenland, with first live photographs and an overview of the class." *Zootaxa* 2035: 41–68.

Museum of Record: Zoological Museum, University of Copenhagen, Copenhagen, Denmark

Obese Diatom

Mann, D. G., S. M. McDonald, M. M. Bayer, S. J. M. Droop, V. A. Chepurnov, R. E. Loke, A. Ciobanu, and J. M. H. du Buf (2004). "The *Sellaphora pupula* species complex (Bacillariophyceae): morphometric analysis, ultrastructure and mating data." *Phycologia* 43(4): 459–482.

Institute of Record: Royal Botanic Garden Edinburgh, Edinburgh ,Scotland, UK

Heckford's Midget Moth

Van Nieukerken, E. J., A. Laštůvka, and Z. Laštůvka (2010). "Western Palaearctic *Ectoedemia* (*Zimmermannia*) Hering and *Ectoedemia* Busck s. str. (Lepidoptera: Nepticulidae): five new species and new data on distribution, hostplants and recognition." *ZooKeys* 32: 1–82.

Museum of Record: Natural History Museum, London, England, UK

Cyprus Mouse

Cucchi, T., A. Orth, J.-C. Auffray, S. Renaud, L. Fabre, J. Catalan, E. Hadjisterkotis, F. Bonhomme, and J.-D. Vigne (2006). "A new endemic species of the subgenus *Mus* (Rodentia, Mammalia) on the Island of Cyprus." *Zootaxa* 1241: 1–36.

Museum of Record: Collection of Vertébrés, Mammifères et Oiseaux, Muséum National d'Histoire Naturelle, Paris, France

Teensiest Chameleon

Glaw, F., J. Köhler, T. M. Townsend, and M. Vences (2012). "Rivaling the world's smallest reptiles: discovery of miniaturized and microendemic new species of leaf chameleons (*Brookesia*) from northern Madagascar." *PLoS ONE* 7(2): e31314, 1–24.

Museum of Record: Zoologische Staatssammlung München, Munich, Germany

Child of Cypris Tiny Fish

Kottelat, M., R. Britz, T. H. Hui, and K.-E. Witte (2006). "*Paedocypris*, a new genus of southeast Asian cyprinid fish with a remarkable sexual dimorphism, comprises the world's smallest vertebrate." *Proceedings of the Royal Society B: Biological Sciences* 273: 895–899.

Museums of Record: Research and Development Centre for Biology, Indonesian Institute of Sciences, Cibinong, Indonesia; Raffles Museum of Biodiversity Research, National University of Singapore, Republic of Singapore

CSIRO's Medusa Jelly

Gershwin, L.-A., and W. Zeidler (2010). "*Csiromedusa medeopolis*: a remarkable Tasmanian medusa (Cnidaria: Hydrozoa: Narcomedusae) comprising a new family, genus and species." *Zootaxa* 2439: 24–34.

Museums of Record: Tasmanian Museum and Art Gallery, Hobart, Tasmania; South Australian Museum, Adelaide, Australia

CHAPTER 4. BIG DEALS: THE LARGEST NEW SPECIES

Golden V Kelp

Kawai, H., T. Hanyuda, M. Lindeberg, and S. C. Lindstrom (2008). "Morphology and molecular phylogeny of *Aureophycus aleuticus* gen. et sp. nov. (Laminariales, Phaeophyceae) from the Aleutian Islands." *Journal of Phycology* 44: 1013–1021.

Museum of Record: University of British Columbia Herbarium, Beaty Biodiversity Museum, Vancouver, British Columbia, Canada

Big Red Jelly

Matsumoto, G. I., K. A. Raskoff, and D. J. Lindsay (2003). "*Tiburonia granrojo* n. sp., a mesopelagic scyphomedusa from the Pacific Ocean representing the type of a new subfamily (class Scyphozoa: order Semaeostomeae: family Ulmaridae: subfamily Tiburoniinae subfam. nov.)." *Marine Biology* 143: 73–77.

Museum of Record: California Academy of Sciences, San Francisco, California, USA

Raptor Fairy Shrimp

Rogers, D. C., D. L. Quinney, J. Weaver, and J. Olesen (2006). "A new giant species of predatory fairy shrimp from Idaho, USA (Branchiopoda: Anostraca)." *Journal of Crustacean Biology*, 26(1): 1–12.

Museum of Record: National Museum of Natural History, Smithsonian Institution, Washington, D.C., USA

Solórzano's Velvet Worm

Morera-Brenes, B., and J. Monge-Nájera (2010). "A new giant species of placented worm and the mechanism by which onychophorans weave their nets (Onychophora: Peripatidae)." *Revista de Biología Tropical/International Journal of Tropical Biology and Conservation* 58(4): 1127–1142.

Museum of Record: Museo de Zoología, Universidad de Costa Rica, San José, Costa Rica

Udzungwa Gray-Faced Sengi

Rovero, F., G. B. Rathbun, A. Perkin, T. Jones, D. O. Ribble, C. Leonard, R. R. Mwakisoma, and N. Doggart. (2008). "A new species of giant sengi or elephant-shrew (genus *Rhynchocyon*) highlights the exceptional biodiversity of the Udzungwa Mountains of Tanzania." *Journal of Zoology* 274: 126–133.

Museum of Record: California Academy of Sciences, San Francisco, California, USA

Huntsman Spider

Jaeger, P. (2001). "A new species of *Heteropoda* (Araneae, Sparassidae, Heteropodinae) from Laos, the largest huntsman spider?" *Zoosystema* 23(3): 461–465.

Museum of Record: Muséum National d'Histoire Naturelle, Paris, France

Chan's Mega-Stick

Hennemann, F. H., and V. O. Conle (2008). "Revision of oriental phasmatodea: the tribe Pharnaciini Günther, 1953, including the description of the world's longest insect, and a survey of the family

Phasmatidae Gray, 1835 with keys to the subfamilies and tribes (Phasmatodea: 'Anareolatae': Phasmatidae)." *Zootaxa* 1906: 1–316.
Museum of Record: Natural History Museum, London, England, UK

Sierra Madre Monitor Lizard

Welton, L. J., C. D. Siler, D. Bennett, A. Diesmos, M. R. Duya, R. Dugay, E. L. B. Rico, M. van Weerd, and R. M. Brown (2010). "A spectacular new Philippine monitor lizard reveals a hidden biogeographic boundary and a novel flagship species for conservation." *Biology Letters* 6(5): 654–658.
Museums of Record: National Museum of the Philippines, Manila, Luzon, Philippines; University of Kansas Natural History Museum, Lawrence, Kansas, USA

Sir Raffles's Showy Flower

Barcelona, J. F., and E. S. Fernando (2002). "A new species of *Rafflesia* (Rafflesiaceae) from Panay Island, Philippines." *Kew Bulletin* 57: 647–651.
Museum of Record: Phillipine National Herbarium, Manila, Philippines

Idip's Starfish

Mah, C. L. (2003). "*Astrosarkus idipi*, A new Indo-Pacific genus and species of Oreasteridae (Valvatida: Asteroidea) displaying extreme endoskeletal reduction." *Bulletin of Marine Science* 73(3): 685–698.
Museum of Record: National Museum of Natural History, Smithsonian Institution, Washington, D.C., USA

CHAPTER 5. SOMETHING OLD, SOMETHING NEW: THE OLDEST NEW SPECIES

Finney's Bat

Simmons, N.B., K. L. Seymour, J. Habersetzer, and G. F. Gunnell (2008). "Primitive Early Eocene bat from Wyoming and the evolution of flight and echolocation." *Nature* 451: 818–821.
Museum of Record: Royal Ontario Museum, Toronto, Ontario, Canada

Aurora Horseshoe Crab

Rudkin, D. M., G. A. Young, and G. S. Nowlan (2008). "The oldest horseshoe crab: a new xiphosurid from Late Ordovician Konservat-Lagerstätten deposits, Manitoba, Canada." *Palaeontology* 51(1): 1–9.
Museum of Record: Manitoba Museum, Winnipeg, Manitoba, Canada

Sarmatian Seahorse

Žalohar, J., T. Hitij, and M. Križnar (2009). "Two new species of seahorses (Syngnathidae, *Hippocampus*) from the Middle Miocene (Sarmatian) Coprolitic Horizon in Tunjice Hills, Slovenia: the oldest fossil record of seahorses." *Annales de Paléontologie* 95: 71–96.

Museum of Record: Zalohar and Hitij Paleontological Collection, Slovenian Museum of Natural History, Ljubljana, Slovena

Sahel Man

Brunet, M., F. Guy, D. Pilbeam, H. T. Mackaye, A. Likius, D. Ahounta, A. Beauvilain, C. Blondel, H. Bocherens, J.-R. Boisserie, L. De Bonis, Y. Coppens, J. Dejax, C. Denys, P. Duringer, V. Eisenmann, G. Fanone, P. Fronty, D. Geraads, T. Lehmann, F. Lihoreau, A. Louchart, A. Mahamat, G. Merceron, G. Mouchelin, O. Otero, P. P. Campomanes, M. Ponce de León, J.-C. Rage, M. Sapanet, M. Schuster, J. Sudre, P. Tassy, X. Valentin, P. Vignaud, L. Viriot, A. Zazzo, and C. Zollikofer (2002). "A new hominid from the Upper Miocene of Chad, Central Africa." *Nature* 418: 145–151.

Museum of Record: Centre National d'Appui à la Recherche, N'Djaména, Chad

Levant Octopus

Fuchs, D., G. Bracchi, and R. Weis (2009). "New octopods (Cephalopoda: Coleoidea) from the Late Cretaceous (Upper Cenomanian) of Hakel and Hadjoula, Lebanon." *Palaeontology* 52(1): 65–81

Museum of Record: Museo Civico di Storia Naturale, Milan, Italy

Burma Bee

Poinar, G. O., Jr., and B. N. Danforth (2006). "A fossil bee from Early Cretaceous Burmese amber." *Science* 314(5799): 614.

Museum of Record: Poinar Amber Collection, Oregon State University, Corvalis, Oregon, USA

Old-Old Mushroom

Poinar, G. O., Jr., and R. Buckley (2007). "Evidence of mycoparasitism and hypermycoparasitism in Early Cretaceous amber." *Mycological Research* 111: 503–506.

Museum of Record: Ron Buckley Amber Collection, Florence, Kentucky, USA

Walking Cactus

Liu, J., M. Steiner, J. A. Dunlop, H. Keupp, D. Shu, Q. Ou, J. Han, Z. Zhang, and X. Zhang (2011). "An armoured Cambrian lobopodian from China with arthropod-like appendages." *Nature* 470: 526–530.

Museum of Record: Early Life Institute, Northwest University, Xi'an, China

Half-Shell Turtle

Li, C., X. Wu, O. Rieppel, L. Wang, and L. Zhao (2008). "An ancestral turtle from the Late Triassic of southwestern China." *Nature* 456: 497–501.

Institution of Record: Institute of Vertebrate Paleontology and Paleoanthropology, Chinese Academy of Sciences, Beijing, China

Dila's Flower

Wang, X., and S. Zheng (2009). "The earliest normal flower from Liaoning Province, China." *Journal of Integrative Plant Biology* 51(8): 800–811.

Museum of Record: Palaeobotanical Collection, Nanjing Institute of Geology and Palaeontology, Chinese Academy of Sciences, Nanjing, China

CHAPTER 6. HELLO, GOOD-BYE: THE MOST ENDANGERED NEW SPECIES

Tennessee Bottlebrush Crayfish

Taylor, C.A., and G. A. Schuster (2010). "Monotypic no more, a description of a new crayfish of the genus *Barbicambarus* Hobbs, 1969 (Decapoda: Cambaridae) from the Tennessee River drainage using morphology and molecules." *Proceedings of the Biological Society of Washington* 123(4): 324–334.

Museums of Record: Illinois Natural History Survey Crustacean Collection, University of Illinois at Urbana-Champaign, Illinois, USA; National Museum of Natural History, Smithsonian Institution, Washington, D.C., USA

Martha's Pink Iguana

Gentile, G., and H. Snell (2009). "*Conolophus marthae* sp. nov. (Squamata, Iguanidae), a new species of land iguana from the Galápagos archipelago." *Zootaxa* 2201: 1–10.

Museum of Record: Civic Museum of Zoology, Rome, Italy

Pygmy Three-Toed Sloth

Anderson, R. P., and C. O. Handley Jr. (2001). "A new species of three-toed sloth (Mammalia: Xenarthra) from Panama, with a review of the genus *Bradypus*." *Proceedings of the Biological Society of Washington* 114(1): 1–33.

Museum of Record: National Museum of Natural History, Smithsonian Institution, Washington, D.C., USA

Cloudy Suckermouth Armored Catfish

Provenzano, F., and N. Milani (2006). "*Cordylancistrus nephelion* (Siluriformes, Loricariidae), a new and endangered species of suckermouth armored catfish from the Tuy River, north-central Venezuela." *Zootaxa* 1116: 29–41.

Museum of Record: Museo de Historia Natural La Salle, Caracas, Venezuela

Parecis Lizard

Colli, G. R., G. C. Costa, A. A. Garda, K. A. Kopp, D. O. Mesquita, A. K. Péres Jr., P. H. Valdujo, G. H. C. Vieira, and H. C. Wiederhecker (2003). "A critically endangered new species of *Cnemidophorus* (Squamata, Teiidae) from a cerrado enclave in southwestern Amazonia, Brazil." *Herpetologica* 59(1): 76–88.

Museum of Record: Coleção Herpetológica da Universidade de Brasília, Brasília, Brazil

Isidoro's Chewing Louse

Perez, J. M., and R. L. Palma (2001). "A new species of *Felicola* (Phthiraptera: Trichodectidae) from the endangered Iberian lynx: another reason to ensure its survival." *Biodiversity and Conservation* 10: 929–937.

Museum of Record: Museo Nacional de Ciencias Naturales, Madrid, Spain

Giant Ribbed Clam

Richter, C., H. Roa-Quiaoit, C. Jantzen, M. Al-Zibdah, and M. Kochzius (2008). "Collapse of a new living species of giant clam in the Red Sea." *Current Biology* 18: 1349–1354.

Museum of Record: Aqaba Marine Science Station, Aqaba, Jordan

Matilda's Horned Viper

Menegon, M., T. R. B. Davenport, and K. M. Howell (2011). "Description of a new and critically endangered species of *Atheris* (Serpentes: Viperidae) from the southern highlands of Tanzania, with an overview of the country's tree viper fauna." *Zootaxa* 3120: 43–54.

Museum of Record: Museo Tridentino di Scienze Naturali, Trento, Italy

Strydom's Yam

Wilkin, P., J. Burrows, S. Burrows, A. M. Muasya, and E. van Wyk (2010). "A critically endangered new species of yam (*Dioscorea strydomiana* Wilkin, Dioscoreaceae) from Mpumalanga, South Africa." *Kew Bulletin* 65: 421–433.

Institution of Record: Pretoria National Botanical Garden, Pretoria, South Africa

Burrunan Dolphin

Charlton-Robb, K., L.-A. Gershwin, R. Thompson, J. Austin, K. Owen, and S. McKechnie (2011). "A new dolphin species, the Burrunan Dolphin *Tursiops australis* sp. nov., endemic to southern Australian coastal waters." *PLoS ONE* 6(9): e24047.

Museum of Record: Queen Victoria Museum and Art Gallery, Launceston, Tasmania, Australia

CHAPTER 7. LETHAL WEAPONS—VENOMS, TOXINS, AND DISEASE: THE DEADLIEST NEW SPECIES

Leprosy Bacterium

Han, X.Y., Y.-H. Seo, K. C. Sizer, T. Schoberle, G. S. May, J. S. Spencer, W. Li, and R. G. Nair (2008). "A new mycobacterium species causing diffuse lepromatous leprosy." *American Journal of Clinical Pathology* 130: 856–864.

Institution of Record: Case 1 DNA sequence at GenBank

Doris Swanson's Poison Dart Frog

Rueda-Almonacid, J. V., M. Rada, S. J. Sánchez-Pacheco, Á. A. Velásquez-Álvarez, and A. Quevedo (2006). "Two new and exceptional poison dart frogs of the genus *Dendrobates* (Anura: Dendrobatidae) from the northeastern flank of the cordillera central of Colombia." *Zootaxa* 1259: 39–54.

Institution of Record: Instituto de Ciencias Naturales-Museo de Historia Natural de la Universidad Nacional de Colombia, Bogotá, Colombia

Corredor's Assassin Bug

Cleber Galvão, C., and V. M. Angulo (2006). "*Belminus corredori*, a new species of Bolboderini (Hemiptera: Reduviidae: Triatominae) from Santander, Colombia." *Zootaxa* 1241: 61–68.

Institution of Record: Herman Lent Collection at the Oswaldo Cruz Institute, Rio de Janeiro, Brazil

Zombie Ant Fungus

Evans, H. C., S. L. Elliot, and D. P. Hughes (2011). "Hidden diversity behind the zombie-ant fungus *Ophiocordyceps unilateralis*: four new species described from carpenter ants in Minas Gerais, Brazil." *PLoS ONE* 6(3): e17024.

Museum of Record: International Mycology Institute at the Royal Botanic Gardens, Kew, London, England, UK

Lilian's Widow Spider

Melic, A. (2000). "The genus *Latrodectus* Walckenaer, 1805 in the Iberian Peninsula (Araneae: Theridiidae) [El género *Latrodectus* walckenaer, 1805 en la Península Ibérica (Araneae: Theridiidae)]." *Revista Ibérica de Aracnología (Iberian Journal of Arachnology)* 1(XII-2000): 13–30.

Museums of Record: Museo Nacional de Ciencias Naturales, Madrid, Spain; Museum National d'Histoire Naturelle, París, France; Natural History Museum, London, England, UK

Ashe's Cobra

Wüster, W., and D. G. Broadley (2007). "Get an eyeful of this: a new species of giant spitting cobra from eastern and north-eastern Africa (Squamata: Serpentes: Elapidae: *Naja*)." *Zootaxa* 1532: 51–68.

Museum of Record: National Museums of Kenya, Nairobi, Kenya

Attenborough's Pitcher

Robinson, A. S., A. S. Fleischmann, S. R. McPherson, V. B. Heinrich, E. P. Gironella, and C. Q. Peña (2009). "A spectacular new species of *Nepenthes* L. (Nepenthaceae) pitcher plant from central Palawan, Philippines." *Botanical Journal of the Linnean Society* 159: 195–202.

Institution of Record: Palawan State University, Puerto Princesa, Palawan, Philippines

Andre Menez's Cone Snail

Biggs, J. S., M. Watkins. P. S. Corneli, and B. M. Olivera (2010). "Defining a clade by morphological, molecular and toxinological criteria: distinctive forms related to *Conus praecellens* A. Adams, 1854." *Nautilus* (Philadelphia), April 6, 2010; 124(1): 1–19.

Institution of Record: Marine Science Institute, University of the Philippines, Quezon City, Philippines

Poisonous Predaceous Polyclad

Ritson-Williams, R., M. Yotsu-Yamashita, and V. J. Paul. (2006). "Ecological functions of tetrodotoxin in a deadly polyclad flatworm." *Proceedings of the National Academy of Sciences* 103(9): 3176–3179.

Institutions of Record: University of Guam Marine Laboratory and the Smithsonian Marine Station at Fort Pierce (experimental sites for feeding and chemical analyses)

King's Deadly Jelly

Gershwin, L.-A. (2007). "*Malo kingi*: A new species of Irukandji jellyfish (Cnidaria: Cubozoa: Carybdeida), possibly lethal to humans, from Queensland, Australia." *Zootaxa* 1659: 55–68.

Museum of Record: Queensland Museum, Brisbane, Australia

CHAPTER 8. GOING TO EXTREMES: NEW SPECIES FROM THE MOST EXTREME ENVIRONMENTS

Cave Pseudoscorpions

Titanobochica magna

Reboleira, A. S. P. S., J. A. Zaragoza, F. Gonçalves, and P. Oromí (2010). "*Titanobochica*, surprising discovery of a new cave-dwelling genus from southern Portugal (Arachnida: Pseudoscorpiones: Bochicidae)." *Zootaxa* 2681: 1–19.

Museum of Record: Departamento de Ecología, Universidad de Alicante, Alicante, Spain

Parobisium yosemite

Cokendolpher, J. C., and J. K. Krejca (2010). "A new cavernicolous *Parobisium* Chamberlin 1930 (Pseudoscorpiones: Neobisiidae) from Yosemite National Park, USA" *Occasional Papers, Museum of Texas Tech University* 297: 1–26.

Museums of Record: Museum of Texas Tech University, Lubbock, Texas, USA; California Academy of Sciences, San Francisco, California, USA; Florida State Collection of Arthropods, Gainesville, Florida, USA

Morafka's Desert Tortoise

Murphy, R. W., K. H. Berry, T. Edwards, A. E. Leviton, A. Lathrop, and J. D. Riedle (2011). "The dazed and confused identity of Agassiz's land tortoise, *Gopherus agassizii* (Testudines, Testudinidae) with the description of a new species, and its consequences for conservation." *ZooKeys* 113: 39–71.

Museum of Record: California Academy of Sciences, San Francisco, California, USA

Yellowstone Bacterium

Bryant, D. A., A. M. Garcia Costas, J. A. Maresca, A. G. M. Chew, C. G. Klatt, M. M. Bateson, L. J. Tallon, J. Hostetler, W. C. Nelson, J. F. Heidelberg, and D. W. Ward (2007). "*Candidatus* Chloracidobacterium thermophilum: an aerobic phototrophic acidobacterium." *Science* 317: 523–526.

Institutions of Record: Pennsylvania State University, University Park, Pennsylvania, USA; Montana State University, Bozeman, Montana, USA

Nature Conservancy Diving Beetle

Miller, K. B., J. R. Gibson, and Y. Alarie (2009). "North American stygobiontic diving beetles (Coleoptera: Dytiscidae: Hydroporinae) with description of *Ereboporus naturaconservatus* Miller, Gibson and Alarie, New Genus and Species, from Texas, U.S.A." *The Coleopterists Bulletin* 63(2): 191–202.

Museum of Record: Museum of Southwestern Biology, University of New Mexico, Albuquerque, New Mexico

Cryptic Forest-Falcon

Whittaker, A. (2002). "A new species of Forest-Falcon (Falconidae: *Micrastur*) from southeastern Amazonia and the Atlantic rainforests of Brazil." *The Wilson Bulletin* 114(4): 421–445.

Museum of Record: Museu Paraense Emílio Goeldi, Belém, Brazil

Stephenson's Antarctic Flower

Rodriguez, E., and P. J. Lopez-Gonzalez (2003). "*Stephanthus antarcticus*, a new genus and species of sea anemone (Actiniaria, Haloclavidae) from the South Shetland Islands, Antarctica." *Helgoland Marine Research* 57: 54–62.

Museums of Record: Zoologisches Institut und Zoologisches Museum Hamburg, Hamburg, Germany; University of Kansas Natural History Museum, Lawrence, Kansas, USA; Zoology Section of the Faculty of Biology, University of Seville, Seville, Spain

Nepalese Autumn Poppy

Egan, P.A. (2011). "*Meconopsis autumnalis* and *M. manasluensis* (Papaveraceae), two new species of Himalayan poppy endemic to central Nepal with sympatric congeners." *Phytotaxa* 20, 47–56.

Museum of Record: Royal Botanic Garden Edinburgh, Scotland, UK

Bare-Faced Bulbul

Woxvold, I. A., J. W. Duckworth, and R. J. Timmins (2009). "An unusual new bulbul (Passeriformes: Pycnonotidae) from the limestone karst of Lao PDR." *Forktail* 25:1–12.

Museums of Record: Natural History Museum, Tring, UK; Australian National Wildlife Collection, Canberra, Australia

Siau Island Tarsier

Shekelle, M., C. Groves, S. Merker, and J. Supriatna (2008). "*Tarsius tumpara*: a new tarsier species from Siau Island, North Sulawesi." *Primate Conservation* (23): 55–64.

Museum of Record: Museum Zoologicum Bogoriense, Bogor, Indonesia

Heat-Loving Tonguefish

Munroe, T. A., and J. Hashimoto (2008). "A new Western Pacific tonguefish (Pleuronectiformes: Cynoglossidae): the first pleuronectiform discovered at active hydrothermal vents." *Zootaxa* 1839: 43–59.

Museum of Record: National Museum of Nature and Science, Tokyo, Japan

CHAPTER 9. THE HIGHEST FORM OF FLATTERY: THE BEST NEW SPECIES MIMICS

Groves's Nudibranch

Hermosillo, A., and Á. Valdés (2008). "Two new species of Opisthobranch mollusks from the tropical eastern Pacific." *Proceedings of the California Academy of Sciences* Series 4, 59(13): 521–532.

Museums of Record: Natural History Museum of Los Angeles County, Los Angeles, California USA; California Academy of Sciences, San Francisco, California, USA

Firefly Flasher

Lloyd, J. E., and L. A. Ballantyne (2003). "Taxonomy and behavior of *Photuris trivittata* sp. n. (Coleoptera: Lampyridae: Photurinae); redescription of *Aspisoma trilineata* (Say) comb. n. (Coleoptera\Lampyridae: Lampyrinae\Cratomorphini)." *Florida Entomologist* 86(4): 464–473.

Museum of Record: James E. Lloyd Collection University of Florida, Gainesville, Florida, USA

Appalachian Tiger Swallowtail

Pavulaan, H., and D. M. Wright (2002). "*Pterourus appalachiensis* (Papilionidae: Papilioninae), a new swallowtail butterfly from the Appalachian region of the United States." *The Taxonomic Report of the International Lepidoptera Survey* 3(7): 1–20.

Museum of Record: Museum of the Hemispheres, Goose Creek, South Carolina, USA

Mache Mountains Glass Frog

Guayasamin, J. M., and E. Bonaccorso (2004). "A new species of glass frog (Centrolenidae: *Cochranella*) from the lowlands of northwestern Ecuador, with comments on the *Cochranella granulosa* group." *Herpetologica* 60(4): 485–494.

Museums of Record: Museo de Zoologia, Universidad Catolica del Ecuador, Quito, Ecuador; University of Kansas Natural History Museum, Lawrence, Kansas, USA

Fish Mime

Schelly, R., T. Takahashi, R. Bills, and M. Hori (2007). "The first case of aggressive mimicry among lamprologines in a new species of *Lepidiolamprologus* (Perciformes: Cichlidae) from Lake Tanganyika." *Zootaxa* 1638: 39–49.

Institution of Record: South African Institute for Aquatic Biodiversity, Grahamstown, South Africa

Ngome Dwarf Chameleon

Tilbury, C. R., and K. A. Tolley (2009). "A new species of dwarf chameleon (Sauria; Chamaeleonidae, *Bradypodion* Fitzinger) from KwaZulu Natal South Africa with notes on recent climatic shifts and their influence on speciation in the genus." *Zootaxa* 2226: 43–57.

Museum of Record: Port Elizabeth Museum, Port Elizabeth, South Africa

Shocking Pink Millipede

Enghoff, H., C. Sutcharit, and S. Panha (2007). "The shocking pink dragon millipede, *Desmoxytes purpurosea*, a colourful new species from Thailand (Diplopoda: Polydesmida: Paradoxosomatidae)." *Zootaxa* 1563: 31–36.

Museums of Record: Museum of Zoology, Chulalongkorn University, Bangkok, Thailand; Natural History Museum of Denmark, University of Copenhagen, Copenhagen, Denmark

Oria's Leaf Insect

Hennemann, F. H., O. V. Conle, M. Gottardo, and J. Bresseel (2009). "On certain species of the genus *Phyllium* Illiger, 1798, with proposals for an intra-generic systematization and the descriptions of five new species from the Philippines and Palawan (Phasmatodea: Phylliidae: Phylliinae: Phylliini)." *Zootaxa* 2322: 1–83.

Museums of Record: Oxford University Museum of Natural History, Oxford, England, UK; Natural History Museum, London, England, UK; Museum für Naturkunde der Humboldt-Universität, Berlin, Germany

Denise's Pygmy Seahorse

Lourie, S. A., and J. E. Randall (2003). "A new pygmy seahorse, *Hippocampus denise* (Teleostei: Syngnathidae), from the Indo-Pacific." *Zoological Studies* 42(2): 284–291.

Museums of Record: Museum Zoologicum Bogoriense, Cibinong, Indonesia; Bishop Museum, Honolulu, Hawaii, USA; Australian Museum, Sydney, Australia

Jumpin' Spider Ant-Mimic

Ceccarelli, F. S. (2010). "New species of ant-mimicking jumping spiders of the genus *Myrmarachne* MacLeay, 1839 (Araneae: Salticidae) from north Queensland, Australia." *Australian Journal of Entomology* 49: 245–255.

Museums of Record: Queensland Museum, Brisbane, Australia; Australian Museum, Sydney, Australia

CHAPTER 10. WHAT'S IN A NAME? NEW SPECIES WITH THE BEST NAMES

Apparating Moon-Gentian

Grant, J. R., and L. Struwe (2003). "*De Macrocarpaeae Grisebach (Ex Gentianaceis) Speciebus Novis III:* six new species of moon-gentians (*Macrocarpaea*, Gentianaceae: Helieae) from Parque Nacional Podocarpus, Ecuador." *Harvard Papers in Botany* 8(1): 61–81.

Museums of Record: National Museum of Natural History, Smithsonian Institution, Washington, D.C., USA; Conservatoire et Jardin botaniques de la Ville de Genève, Geneva, Switzerland

Madonna's Water Bear

Michalczyk, Ł., and Ł. Kaczmarek (2006). "Revision of the *Echiniscus bigranulatus* group with a description of a new species *Echiniscus madonnae* (Tardigrada: Heterotardigrada: Echiniscidae) from South America." *Zootaxa* 1154: 1–26.

Museum of Record: Natural Sciences Collection, Adam Mickiewicz University, Poznań, Poland

Bonaire Banded Box Jelly

Collins, A. G., B. Bentlage, W. Gillan, T. H. Lynn, A. C. Morandini, A. C. Marques (2011). "Naming the Bonaire banded box jelly, *Tamoya ohboya*, n. sp. (Cnidaria: Cubozoa: Carybdeida: Tamoyidae." *Zootaxa* 2753: 53–68.

Museum of Record: National Museum of Natural History, Smithsonian Institution, Washington, D.C., USA

John Cleese's Woolly Lemur

Thalmann, U., and T. Geissmann (2005). "New species of woolly lemur *Avahi* (Primates: Lemuriformes) in Bemaraha (Central Western Madagascar)." *American Journal of Primatology* 67: 371–376.

Museum of Record: hair samples at the Anthropological Institute and Museum, University of Zurich, Zurich, Switzerland

Google Ant

Fisher, B. L. (2005). "A new species of *Discothyrea* Roger from Mauritius and a new species of *Proceratium* Roger from Madagascar (Hymenoptera: Formicidae)." *Proceedings of the California Academy of Sciences* 56(35): 657–667.

Museums of Record: California Academy of Sciences, San Francisco, California, USA; Museum of Comparative Zoology, Harvard University, Cambridge, Massachusetts, USA

David Bowie's Spider

Jäger, P. (2008). "Revision of the huntsman spider genus *Heteropoda* Latreille 1804: species with exceptional male palpal conformations (Araneae: Sparassidae: Heteropodinae)." *Senckenbergiana biologica* 88(2): 239–310.

Museums of Record: Research Institute Senckenberg, Frankfurt am Main, Germany; Raffles Museum of Biodiversity Research, Singapore, Republic of Singapore

SpongeBob SquarePants Fungus

Desjardin, D. E., K. G. Peay, and T. D. Bruns (2011). "*Spongiforma squarepantsii*, a new species of gasteroid bolete from Borneo." *Mycologia* 103(5): 1119–1123.

Institutions of Record: University of California, Berkeley, California; Forest Department, Sarawak, Malaysia

Clare Hannah's Shrimp

McCallum, A. W., and G. C. B. Poore (2010). "Two crested and colourful new species of *Lebbeus* (Crustacea: Caridea: Hippolytidae) from the continental margin of Western Australia." *Zootaxa* 2372: 126–137.

Museum of Record: Western Australia Museum, Perth, Australia

Groening's Sand Crab

Boyko, C. B. (2002). "A worldwide revision of the recent and fossil sand crabs of the Albuneidae Stimpson and Blepharipodidae, new family (Crustacea: Decapoda: Anomura: Hippoidea)." *Bulletin of the American Museum of Natural History* 272: 1–396.

Museums of Record: Rijksmuseum van Natuurlijke Historie (now Nationaal Natuurhistorisch Museum), Leiden, the Netherlands; Natural History Museum, London, England, UK

Wonderfully Photogenic 'Pus

Hochberg, F. G., M. D. Norman, and J. Finn (2006). "*Wunderpus photogenicus* n. gen. and sp., a new octopus from the shallow waters of the Indo-Malayan Archipelago (Cephalopoda: Octopodidae)." *Molluscan Research* 26(3): 128–140.

Museums of Record: Australian Museum, Sydney, Australia; Santa Barbara Museum of Natural History, Santa Barbara, California, USA

Index

Note: Page numbers in *italics* refer to maps or illustrations.

adaptations, 26
Albunea groeningi (sand crab), *216*, 236–37
Aleutian Islands: *Aureophycus aleuticus* (kelp), *72*, 76–77
algae: *Sellaphora obesa* (diatom), *48*, 60–61
Amanita phalloides (death cap mushroom), *146*
amphibians
  Cochranella mache (frog), *192*, 202–3
  Dendrobates dorisswansoni (frog), *144*, 146, 150–51
  Nasikabatrachus sahyadrensis (frog), *24*, 44–45
anemones: *Stephanthus antarcticus*, *168*, 182–83
Antarctica: *Stephanthus antarcticus* (anemone), *168*, 182–83
ants
  Myrmarachne smaragdina (ant mimic spider), 192, 214–15
  Ophiocordyceps camponoti-rufipedis (zombie ant fungus), *144*, 154–55
  Proceratium google (ant), *216*, 228–29
aposematism, 5, 193
arachnids. *See* spiders
Archaefructus (flower), 119
Archipheretima middletoni (worm), *xxvi*, 14–15
Arcticotantulus kristenseni (Tantulocaridan), *48*, 58–59
Arcticotantulus pertzovi (Tantulocaridan), 59
assassins
  Belminus corredori (assassin bugs), *144*, 152–53
  Eriauchenius lavatenda (assassin spider), *24*, 42–43
Astrosarkus idipi (starfish), *72*, 94–95
Atheris matildae (snake), *120*, 138–39
Atlantic Ocean
  Grimpoteuthis discoveryi (Dumbo octopus), *24*, 36–37

Lasiognathus amphirhamphus (anglerfish), *24*, 34–35
Reticulammina cerebreformis (protist), *24*, 38–39
Attenborough, Sir David, 49
Aureophycus aleuticus (kelp), *72*, 76–77
Australia
Lebbeus clarehannah (shrimp), *216*, 234–35
Malo kingi (jellyfish), *144*, 166–67
Myrmarachne smaragdina (spider), *192*, 214–15
Tursiops australis (dolphin), *120*, 142–43
Australopithecus afarensis (Lucy), 107
Ausubel's Mighty Claw Lobster, *24*, 46–47
Avahi cleesei (woolly lemur), *216*, 226–27

bacteria
Candidatus Chloracidobacterium thermophilum, *168*, 176–77
Mycobacterium lepromatosis, *144*, 148–49
bamboo: *Raddiella vanessiae*, *48*, 54–55
Barbicambarus cornutus (crayfish), 125
Barbicambarus simmonsi (crayfish), *120*, 124–25
Barbie Pagoda Fungus, *xxvi*, 20–21
Batesian mimicry, 193–94
bats
Onychonycteris finneyi, *96*, 100–101
Platyrrhinus masu, *24*, 30–31
bees
Euglossa paisa, *xxvi*, 6–7
Melittosphex burmensis, *96*, 110–11
beetles: *Ereboporus naturaconservatus*, *168*, 178–79
Belminus corredori (assassin bugs), *144*, 152–53
Berthella grovesi (nudibranch), *192*, 196–97
biodiversity, xxi–xxii, 26–27, 121–22
birds
Glaucidium mooreorum (pygmy-owl), *48*, 56–57
Kelenken guillermoi (terror bird), 74
Micrastur mintoni (forest falcon), *168*, 180–81
Pycnonotus hualon (songbird), *168*, 186–87
Borneo
Spongiforma squarepantsii (fungus), *216*, 232–33
Tarsius tumpara (tarsiers), *168*, 188–89
Bradypodion ngomeense (chameleon), *192*, 206–7
Bradypus pygmaeus (sloth), *120*, 128–29
Branchinecta raptor (shrimp), *72*, 80–81
Brazil, 122
Cnemidophorus parecis (lizard), *120*, 132–33
Coendou roosmalenorum (porcupine), *48*, 52–53
Glaucidium mooreorum (pygmy-owl), *48*, 56–57
Heliotrygon gomesi (stingray), *24*, 32–33
Micrastur mintoni (forest falcon), *168*, 180–81
Oligoxystre diamantinensis (tarantula), *xxvi*, 10–11
Ophiocordyceps camponoti-rufipedis (fungus), *144*, 154–55
Brookesia micra (chameleon), *48*, 66–67
Burma (Myanmar)
Melittosphex burmensis (bee), *96*, 110–11
Palaeoagaracites antiquus (mushroom), *96*, 112–13

butterflies: *Pterourus appalachiensis*, *192*, 200–201
Caerostris darwini (spider), 74
camouflage, 193
Canada: *Lunataspis aurora* (horseshoe crab), *96*, 102–3
Candidatus Chloracidobacterium thermophilum (bacterium), *168*, 176–77
carnivorous plants: *Nepenthes attenboroughii*, *144*, 160–61
catfish: *Cordylancistrus nephelion*, *120*, 130–31
Chad: *Sahelanthropus tchadensis* (hominid), *96*, 106–7
Chagas' disease, 153
chameleons
Bradypodion ngomeense (chameleon), *192*, 206–7
Brookesia micra (chameleon), *48*, 66–67
Chan's mega-stick, *72*, 88–89
China
Callianthus dilae (flower), *96*, 118–19
Diania cactiformis (lobopodian), *96*, 114–15
Odontochelys semitestacea (turtle), *96*, 116–17
chromosome numbers, multiplication of, xix
cladistic analysis, xviii, xx
clams: *Tridacna costata*, *120*, 136–37
Cnemidophorus parecis (lizard), *120*, 132–33
cobras: *Naja ashei*, *144*, 158–59
Cochranella mache (frog), *192*, 202–3
Coelopleurus exquisitus (sea urchin), *xxvi*, 22–23
Coendou roosmalenorum (porcupine), *48*, 52–53
Colombia
Belminus corredori (assassin bugs), *144*, 152–53
Dendrobates dorisswansoni (frog), *144*, 150–51
Euglossa paisa (Paisa orchid bee), *xxvi*, 6–7
complexity, trends in, 50–51
Condylago furculifera (orchid), *24*, 28–29
Conolophus marthae (iguana), *120*, 126–27
Conus andremenezi (snail), *144*, 162–63
Conus geographus (snail), 163
Cordylancistrus nephelion (catfish), *120*, 130–31
corpse flower (*Rafflesia speciosa*), *72*, 92–93
Costa Rica: *Peripatus solorzanoi* (worm), *72*, 82–83
crabs: *Albunea groeningi* (sand crab), *216*, 236–37
crayfish
Barbicambarus cornutus, *125*
Barbicambarus simmonsi, *120*, 124–25
Cretaceous-Tertiary (K-T) boundary event, 98
crustaceans
Albunea groeningi (sand crab), *216*, 236–37
Arcticotantulus kristenseni (Tantulocaridan), *48*, 58–59
Barbicambarus cornutus (crayfish), 125
Barbicambarus simmonsi (crayfish), *120*, 124–25
Branchinecta raptor (shrimp), *72*, 80–81
Dinochelus ausubeli (Ausubel's mighty claw lobster), *24*, 46–47
Lebbeus clarehannah (shrimp), *216*, 234–35
Cryptogreagris steinmanni (pseudoscorpion), 173
Csiromedusa medeopolis (jellyfish), *48*, 70–71
curiosity of humans, xxi

cuscuses: *Spilocuscus wilsoni* (blue-eyed cuscus), *xxvi*, 18–19
Cyprus: *Mus cypriacus* (mouse), xviii, *48*, 64–65

Darwin, Charles, xx, 127
deadliest new species, *144*, 145–167
death cap mushroom (*Amanita phalloides*), *146*
Dendrobates dorisswansoni (frog), *144*, 150–51
Desmoxytes purpurosea (millipede), *192*, 208–9
devil's worm (*Halicephalobus mephisto*), 170–71
diatoms: *Sellaphora obesa*, *48*, 60–61
Dinochelus ausubeli (Ausubel's mighty claw lobster), *24*, 46–47
Dioscorea strydomiana (yam), *120*, 140–41
discovery of new species, xvii, xix–xx, xxiii
Disko Bay and Disko Fjord, off coast of Greenland: *Arcticotantulus kristenseni* (Tantulocaridan), *48*, 58–59
dolphins: *Tursiops australis*, *120*, 142–43
Dumbo octopus, *24*, 36–37

earthstars, *24*, 40–41
Echiniscus madonnae (tardigrade), *216*, 222–23
Ectoedemia heckfordi (moth), *48*, 62–63
Ecuador
  Cochranella mache (frog), *192*, 202–3
  Conolophus marthae (iguana), *120*, 126–27
  Macrocarpaea apparata (flowering tree), *216*, 220–21
elephant shrews (sengi), *72*, 84–85
endangered new species, *120*, 121–143
Entropezites patricii (fungus), 113
Ereboporus naturaconservatus (beetle), *168*, 178–79
Eriauchenius lavatenda (spider), *24*, 42–43
Euglossa paisa (Paisa orchid bee), *xxvi*, 6–7
evolution, xviii–xix, xx–xxi, 26
Exquisite Sea Urchin, 22–23
extinction, 98, 121–22, 135
extreme environments, new species from, *168*, 169–191

Felicola isidoroi (lice), *120*, 134–35
fireflies: *Photuris trivittata*, *192*, 198–99
fish
  Cordylancistrus nephelion (catfish), *120*, 130–31
  Heliotrygon gomesi (stingray), *24*, 32–33
  Hippoçampus (seahorse), *96*, 104–105, *192*, 212–13
  Histiophryne psychedelica (psychedelic frogfish), *xxvi*, 16–17
  Lasiognathus amphirhamphus (anglerfish), *24*, 34–35
  Lepidiolamprologus mimicus (cichlid fish), *192*, 204–5
  Paedocypris progenetica, *48*, 68–69
  puffer fish, 165
  Schindleria brevipinguis, 69
  Symphurus thermophilus (flatfish), *168*, 190–91
flatworms: "Planocerid species 1," *144*, 164–65
flowering trees: *Macrocarpaea apparata*, *216*, 220–21
flowers
  Callianthus dilae (Dila's flower), *96*, 118–19

Condylago furculifera (orchid), *24*, 28–29
Meconopsis autumnalis (poppies), *168*, 184–85
Phragmipedium kovachii (orchid), *xxvi*, 8–9
Rafflesia speciosa (corpse flower), *72*, 92–93
fossil species, xxii, *96*, 97–119
French Guiana: *Raddiella vanessiae* (bamboo), *48*, 54–55
frogfish, *xxvi*, 16–17
frogs
Cochranella mache, *192*, 202–3
Dendrobates dorisswansoni, *144*, 146, 150–51
Nasikabatrachus sahyadrensis, *24*, 44–45
fungi
Amanita phalloides, *146*
Entropezites patricii, *113*
Geastrum episcopale, 41
Geastrum parvistriatum, *24*, 40–41
Lycoperdon hyemale, *41*
Mycetophagites atrebora, 113
Ophiocordyceps camponoti-rufipedis, *144*, 154–55
Palaeoagaracites antiquus, *96*, 112–13
Podoserpula miranda, *xxvi*, 20–21
Spongiforma squarepantsii, *216*, 232–33

Galápagos Islands, 127
Geastrum episcopale (earthstar), 41
Geastrum parvistriatum (earthstar fungi), *24*, 40–41
geographic cone snail (*Conus geographus*), 163
Glaucidium mooreorum (pygmy-owl), *48*, 56–57
Goliath beetles, 50
Gopherus morafkai (tortoise), *168*, 174–75
Grimpoteuthis discoveryi (Dumbo octopus), *24*, 36–37
Guam: "Planocerid species 1" (flatworm), *144*, 164–65

habitat loss, 121–23
Haeckel, Ernst, *xix*
Halicephalobus mephisto (devil's worm), 170–71
Hansen's disease (leprosy), 149
Heckford's Midget Moth, *48*, 62–63
Heliotrygon gomesi (stingray), *24*, 32–33
Heteropoda davidbowie (spider), *216*, 230–31
Heteropoda maxima (spider), *72*, 86–87
Hippocampus denise (seahorse), *192*, 212–13
Hippocampus slovenicus (seahorse), 105
Histiophryne psychedelica (psychedelic frogfish), *xxvi*, 16–17
hominids
Australopithecus afarensis (Lucy), 107
Orrorin tugenensis (Millennium Man), 107
Sahelanthropus tchadensis (hominid), *96*, 106–7
horseshoe crabs: *Lunataspis aurora*, *96*, 102–3
host specificity, 135
human ancestors. *See* hominids

Iberian lynx (*Lynx pardinus*), 135
iguanas: *Conolophus marthae*, *120*, 126–27
India: *Nasikabatrachus sahyadrensis* (frog), *24*, 44–45
Indonesia
Hippocampus denise (seahorse), *192*, 212–13

Histiophryne psychedelica (psychedelic frogfish), *xxvi*, 16–17
Paedocypris progenetica (fish), *48*, 68–69
Spilocuscus wilsoni (blue-eyed cuscus), *xxvi*, 18–19
Tarsius tumpara (tarsiers), *168*, 188–89
Wunderpus photogenicus (octopus), *216*, 238–39
insects, 49, 50
Belminus corredori (assassin bugs), *144*, 152–53
Ectoedemia heckfordi (moth), *48*, 62–63
Ereboporus naturaconservatus (beetle), *168*, 178–79
Euglossa paisa (Paisa orchid bee), *xxvi*, 6–7
Felicola isidoroi (lice), *120*, 134–35
Melittosphex burmensis (bee), *96*, 110–11
Phobaeticus chani (stick insect), *72*, 88–89
Photuris trivittata (firefly), *192*, 198–99
Phyllium ericoriai (leaf insect), *192*, 210–11
Proceratium google (ant), *216*, 228–29
Pterourus appalachiensis (butterfly), *192*, 200–201

Japan: *Albunea groeningi* (sand crab), *216*, 236–37
jellyfish
Csiromedusa medeopolis, *48*, 70–71
Malo kingi, *144*, 166–67
Tamoya ohboya, *216*, 224–25
Tiburonia granrojo, *72*, 78–79

Kelenken guillermoi (terror bird), 74
kelp: *Aureophycus aleuticus*, *72*, 76–77
Kenya: *Naja ashei* (cobra), *144*, 158–59
Keuppia hyperbolaris (octopus), 109
Keuppia levante (octopus), *96*, 108–9

Lao People's Democratic Republic (Laos)
Heteropoda maxima (spider), *72*, 86–87
Pycnonotus hualon (bird), *168*, 186–87
largest new species, *72*, 73–95
Lasiognathus amphirhamphus (anglerfish), *24*, 34–35
Latrodectus lilianae (spider), *144*, 156–57
leaf insect: *Phyllium ericoriai*, *192*, 210–11
Lebanon: *Keuppia levante* (octopus), *96*, 108–9
Lebbeus clarehannah (shrimp), *216*, 234–35
lemurs: *Avahi cleesei* (woolly lemur), *216*, 226–27
Lepidiolamprologus mimicus (fish), *192*, 204–5
lethal new species, *144*, 145–167
lice: *Felicola isidoroi*, *120*, 134–35
Linné, Carl von "Linnaeus," xvii, *xviii*
Liophidium pattoni (snake), *xxvi*, 12–13
Little Fork Orchid, 28–29
lizards
Bradypodion ngomeense (chameleon), *192*, 206–7
Brookesia micra (chameleon), *48*, 66–67
Cnemidophorus parecis, *120*, 132–33
Conolophus marthae (iguana), *120*, 126–27
Varanus bitatawa (monitor lizard), *72*, 90–91
Lobopodia, 115
lobopodians: *Diania cactiformis*, *96*, 114–15
lobsters: *Dinochelus ausubeli* (Ausubel's mighty claw lobster), *24*, 46–47
Lucy (*Australopithecus afarensis*), 107

Lunataspis aurora (horseshoe crab), *96*, 102–3
Lycoperdon hyemale (fungus), *41*
Lynx pardinus (Iberian lynx), 135

Macrocarpaea apparata (flowering tree), *216*, 220–21
Madagascar, 122
Avahi cleesei (woolly lemur), *216*, 226–27
Brookesia micra (chameleon), *48*, 66–67
Eriauchenius lavatenda (spider), *24*, 42–43
Liophidium pattoni (snake), *xxvi*, 12–13
Proceratium google (ant), *216*, 228–29
Malaysia
Heteropoda davidbowie (spider), *216*, 230–31
Phobaeticus chani (stick insect), *72*, 88–89
Spongiforma squarepantsii (fungus), *216*, 232–33
Malo kingi (jellyfish), *144*, 166–67
mammals
Avahi cleesei (woolly lemur), *216*, 226–27
Bradypus pygmaeus (sloth), *120*, 128–29
Coendou roosmalenorum (porcupine), *48*, 52–53
Lynx pardinus (Iberian lynx), 135
Mus cypriacus (mouse), xviii, *48*, 64–65
Onychonycteris finneyi (bat), *96*, 100–101
Platyrrhinus masu (bat), *24*, 30–31
Rhynchocyon udzungwensis (sengi), *72*, 84–85
Spilocuscus wilsoni (blue-eyed cuscus), *xxvi*, 18–19
Tursiops australis (dolphin), *120*, 142–43
marine animals
Astrosarkus idipi (starfish), *72*, 94–95
Coelopleurus exquisitus (sea urchin), *xxvi*, 22–23
Csiromedusa medeopolis (jellyfish), *48*, 70–71
Grimpoteuthis discoveryi (Dumbo octopus), *24*, 36–37
Hippocampus denise (seahorse), *192*, 212–13
Hippocampus sarmaticus (seahorse), *96*, 104–5
Keuppia hyperbolaris (octopus), 109
Keuppia levante (octopus), *96*, 108–9
Malo kingi (jellyfish), *144*, 166–67
Styletoctopus annae (octopus), 109
Tamoya ohboya (jellyfish), *216*, 224–25
Tiburonia granrojo (jellyfish), *72*, 78–79
Tridacna costata (clam), *120*, 136–37
Tursiops australis (dolphin), *120*, 142–43
Wunderpus photogenicus (octopus), *216*, 238–39
See also crustaceans; fish
marsupials: *Spilocuscus wilsoni* (blue-eyed cuscus), *xxvi*, 18–19
Meconopsis autumnalis (poppies), *168*, 184–85
Melittosphex burmensis (bee), *96*, 110–11
Mexico
Berthella grovesi (nudibranch), *192*, 196–97
Photuris trivittata (firefly), *192*, 198–99
Polycera kaiserae (nudibranch), *xxvi*, 4–5
mice: *Mus cypriacus*, *48*, 64–65
Micrastur mintoni (forest falcon), *168*, 180–81
microbial species, 49–50
Millennium Man (*Orrorin tugenensis*), 107
millipedes: *Desmoxytes purpurosea*, *192*, 208–9
mimics among new species, *192*, 193–215
miniaturization, *48*, 49–71

mollusks
  Conus andremenezi (snail), *144*, 162–63
  Conus geographus (snail), 163
  Grimpoteuthis discoveryi (octopus), *24*, 36–37
  Keuppia hyperbolaris (octopus), 109
  Keuppia levante (octopus), *96*, 108–9
  Styletoctopus annae (octopus), 109
  Tridacna costata (clam), *120*, 136–37
  Wunderpus photogenicus (octopus), *216*, 238–39
monitor lizards: *Varanus bitatawa*, *72*, 90–91
moths: *Ectoedemia heckfordi*, *48*, 62–63
Müllerian mimicry, 194
Mus cypriacus (mouse), *48*, 64–65
mushrooms
  Amanita phalloides, *146*
  Palaeoagaracites antiquus, *96*, 112–13
  See also fungi
Myanmar (Burma)
  Melittosphex burmensis (bee), *96*, 110–11
  Palaeoagaracites antiquus (mushroom), *96*, 112–13
Mycetophagites atrebora (fungus), 113
Mycobacterium lepromatosis (bacterium), *144*, 148–49
Myrmarachne smaragdina (spider), *192*, 214–15

Naja ashei (snake), *144*, 158–59
names (best) of new species, *216*, 217–239
Nasikabatrachus sahyadrensis (frog), *24*, 44–45
natural selection, xx, 26
Nepal: *Meconopsis autumnalis* (poppies), *168*, 184–85
Nepenthes attenboroughii (carnivorous plant), *144*, 160–61
New Caledonia
  Coelopleurus exquisitus (sea urchin), *xxvi*, 22–23
  Podoserpula miranda (Barbie Pagoda Fungus), *xxvi*, 20–21
new species, defined, xviii
Nosy Hara chameleon, *48*, 66–67
nudibranchs
  Berthella grovesi (nudibranch), *192*, 196–97
  Polycera kaiserae (nudibranch), *xxvi*, 4–5

octopuses
  Grimpoteuthis discoveryi, *24*, 36–37
  Keuppia hyperbolaris, *109*
  Keuppia levante, *96*, 108–9
  Styletoctopus annae, 109
  Wunderpus photogenicus, *216*, 238–39
oldest new species, *96*, 97–119
Oligoxystre diamantinensis (tarantula), *xxvi*, 10–11
Onychonycteris finneyi (bat), *96*, 100–1
onychophorans (velvet worms), *72*, 82–83
Ophiocordyceps camponoti-rufipedis (fungus), *144*, 154–55
orchid bees, 6–7
orchids
  Condylago furculifera (orchid), *24*, 28–29
  and orchid bees, 6–7
  Phragmipedium kovachii (orchid), *xxvi*, 8–9
Orrorin tugenensis (Millennium Man), 107
owls: *Glaucidium mooreorum* (pygmy-owl), *48*, 56–57

Pacific Ocean
Astrosarkus idipi (starfish), *72*, 94–95
Symphurus thermophilus (fish), *168*, 190–91
Tiburonia granrojo (jellyfish), *72*, 78–79
Wunderpus photogenicus (octopus), *216*, 238–39
Paedocypris progenetica (fish), *48*, 68–69
Paisa orchid bee, *xxvi*, 6–7
Palau: *Astrosarkus idipi* (starfish), *72*, 94–95
paleontology, 97–98. *See also* fossil species
Panama
Bradypus pygmaeus (sloth), *120*, 128–29
Condylago furculifera (orchid), *24*, 28–29
Parobisium yosemite (pseudoscorpion), *168*, 172–73
Peckhamian mimicry, 194
pentamerism, 23
Peripatus solorzanoi (worm), *72*, 82–83
Peru
Echiniscus madonnae (tardigrade), *216*, 222–23
Phragmipedium kovachii (orchid), *xxvi*, 8–9
Platyrrhinus masu (broad-nosed bat), *24*, 30–31
phasmids: *Phyllium ericoriai* (leaf insect), *192*, 210–11
Philippines
Albunea groeningi (sand crab), *216*, 236–37
Archipheretima middletoni (worm), *xxvi*, 14–15
Conus andremenezi (snail), *144*, 162–63
Nepenthes attenboroughii (carnivorous plant), *144*, 160–61
Phyllium ericoriai (leaf insect), *192*, 210–11
Rafflesia speciosa (flower), *72*, 92–93
Varanus bitatawa (monitor lizard), *72*, 90–91
Philippine Sea: *Dinochelus ausubeli* (Ausubel's mighty claw lobster), *24*, 46–47
Phobaeticus chani (stick insect), *72*, 73, 88–89
Photuris trivittata (firefly), *192*, 198–99
Phragmipedium kovachii (orchid), *xxvi*, 8–9
Phyllium ericoriai (leaf insect), *192*, 210–11
pitcher plants (*Nepenthes attenboroughii*), *144*, 160–61
"Planocerid species 1" (flatworm), *144*, 164–65
plants
Archaefructus (flower), 119
Callianthus dilae (Dila's flower), *96*, 118–19
Condylago furculifera (orchid), *24*, 28–29
Dioscorea strydomiana (yam), *120*, 140–41
Macrocarpaea apparata (flowering tree), *216*, 220–21
Meconopsis autumnalis (poppies), *168*, 184–85
mimicry of, 194–95
Nepenthes attenboroughii (carnivorous plant), *144*, 160–61
Phragmipedium kovachii (orchid), *xxvi*, 8–9
Raddiella vanessiae (bamboo), *48*, 54–55
Rafflesia speciosa (corpse flower), *72*, 92–93
See also fungi
Platyrrhinus masu (broad-nosed bat), *24*, 30–31
Podoserpula miranda (Barbie pagoda fungus), *xxvi*, 20–21
poison dart frogs (*Dendrobates dorisswansoni*), *144*, 146, 150–51
Polycera kaiserae (nudibranch), *xxvi*, 4–5

polyploidy, xix
porcupines: *Coendou roosmalenorum*, *48*, 52–53
Portugal: *Titanobochica magna* (pseudoscorpion), *168*, 172–73
possums, *xxvi*, 18–19
prettiest species, *xxvi*, 1–23
Proceratium google (ant), *216*, 228–29
Proganochelys (turtle), 117
protists: *Reticulammina cerebreformis*, *24*, 38–39
  protozoa, *51*
pseudoscorpions, *168*, 172–73
psychedelic frogfish, *xxvi*, 16–17
Pterourus appalachiensis (butterfly), *192*, 200–201
puffer fish, 165
Pycnonotus hualon (bird), *168*, 186–87

Raddiella vanessiae (bamboo), *48*, 54–55
Rafflesia speciosa (flower), *72*, 92–93
Red Sea: *Tridacna costata* (clam), *120*, 136–37
reproduction, xix
reptiles
  Atheris matildae (snake), *120*, 138–39
  Bradypodion ngomeense (chameleon), *192*, 206–7
  Brookesia micra (chameleon), *48*, 66–67
  Cnemidophorus parecis, *120*, 132–33
  Conolophus marthae (iguana), *120*, 126–27
  Gopherus morafkai (tortoise), *168*, 174–75
  Odontochelys semitestacea (turtle), *96*, 116–17
  Proganochelys (turtle), 117
  Varanus bitatawa (monitor lizard), *72*, 90–91
Reticulammina cerebreformis (protist), *24*, 38–39
Rhynchocyon udzungwensis (sengi), *72*, 84–85

Sahelanthropus tchadensis (Toumaë), *96*, 106–7
Schindleria brevipinguis (fish), 69
seahorses
  Hippocampus denise (seahorse), *192*, 212–13
  Hippocampus sarmaticus (seahorse), *96*, 104–5
  Hippocampus slovenicus (seahorse), 105
sea slugs (nudibranchs)
  Berthella grovesi (nudibranch), *192*, 196–97
  Polycera kaiserae (nudibranch), *xxvi*, 4–5
sea stars: *Astrosarkus idipi*, *72*, 94–95
sea urchins: *Coelopleurus exquisitus*, *xxvi*, 22–23
sea vents, hydrothermal, 170, 191
Sellaphora obesa (diatom), *48*, 60–61
sengi: *Rhynchocyon udzungwensis*, *72*, 84–85
shrimp
  Branchinecta raptor (shrimp), *72*, 80–81
  Lebbeus clarehannah (shrimp), *216*, 234–35
sloths: *Bradypus pygmaeus* (sloth), *120*, 128–29
Slovenia: *Hippocampus sarmaticus* (seahorse), *96*, 104–5
smallest new species, *48*, 49–71
snails
  Conus andremenezi, *144*, 162–63
  Conus geographus, *163*
snakes
  Atheris matildae, *120*, 138–39

Liophidium pattoni, *xxvi*, 12–13
Naja ashei, *144*, 158–59
South Africa
Bradypodion ngomeense (chameleon), *192*, 206–7
Dioscorea strydomiana (yam), *120*, 140–41
Spain
Felicola isidoroi (lice), *120*, 134–35
Geastrum parvistriatum (fungi), *24*, 40–41
Latrodectus lilianae (spider), *144*, 156–57
speciation, xviii–xix
spiders
Caerostris darwini, 74
Eriauchenius lavatenda, *24*, 42–43
Heteropoda davidbowie, *216*, 230–31
Heteropoda maxima, *72*, 86–87
Latrodectus lilianae, *144*, 156–57
Myrmarachne smaragdina, *192*, 214–15
Oligoxystre diamantinensis, *xxvi*, 10–11
Spilocuscus wilsoni (blue-eyed cuscus), *xxvi*, 18–19
Spongiforma squarepantsii (fungus), *216*, 232–33
starfish: *Astrosarkus idipi*, *72*, 94–95
Stephanthus antarcticus (anemone), *168*, 182–83
stick insects: *Phobaeticus chani*, *72*, 88–89
stingrays: *Heliotrygon gomesi*, *24*, 32–33
strangest new species, *24*, 25–47
Styletoctopus annae (octopus), 109
Sulawesi: *Tarsius tumpara* (tarsiers), *168*, 188–89
sustainability, xxi–xxii
Symphurus thermophilus (fish), *168*, 190–91
Tamoya ohboya (jellyfish), *216*, 224–25
Tantulacus dieteri (Tantulocaridan), 59
Tantulacus karolae (Tantulocaridan), 59
Tantulocaridans: *Arcticotantulus kristenseni*, *48*, 58–59
Tanzania
Atheris matildae (snake), *120*, 138–39
Rhynchocyon udzungwensis (sengi), *72*, 84–85
tarantulas: *Oligoxystre diamantinensis*, *xxvi*, 10–11
tardigrades: *Echiniscus madonnae*, *216*, 222–23
Tarsius tumpara (tarsiers), *168*, 188–89
Tasmania: *Csiromedusa medeopolis* (jellyfish), *48*, 70–71
taxonomy, xvii–xviii, xix, xx, xxiii
telomerase, 47
terror bird (*Kelenken guillermoi*), 74
tetrodotoxin, 165
Thailand: *Desmoxytes purpurosea* (millipede), *192*, 208–9
thelytoky, xix
Tiburonia granrojo (jellyfish), *72*, 78–79
Titanobochica magna (pseudoscorpion), *168*, 172–73
trees: *Macrocarpaea apparata*, *216*, 220–21
Tridacna costata (clam), *120*, 136–37
Tursiops australis (dolphin), *120*, 142–43
turtles and tortoises
Gopherus morafkai (tortoise), *168*, 174–75
Odontochelys semitestacea (turtle), *96*, 116–17
Proganochelys (turtle), 117

United Kingdom
Ectoedemia heckfordi (moth), *48*, 62–63
Sellaphora obesa (diatom), *48*, 60–61
United States
Barbicambarus simmonsi (crayfish), *120*, 124–25
Branchinecta raptor (shrimp), *72*, 80–81
Chloracidobacterium thermophilum (bacterium), *168*, 176–77
Ereboporus naturaconservatus (beetle), *168*, 178–79
Gopherus morafkai (tortoise), *168*, 174–75
Mycobacterium lepromatosis (bacterium), *144*, 148–49
Onychonycteris finneyi (bat), *96*, 100–101
Pterourus appalachiensis (butterfly), *192*, 200–201

Varanus bitatawa (monitor lizard), *72*, 90–91
velvet worms (onychophorans), *72*, 82–83
Venezuela: *Cordylancistrus nephelion* (catfish), *120*, 130–31

walking sticks: *Phobaeticus chani* (stick insect), *72*, *73*, 88–89
wasps, 111
widow spiders, *144*, 156–57
wildcats: *Lynx pardinus* (Iberian lynx), 135
Wilson, E. O., 49
woolly lemurs: *Avahi cleesei*, *216*, 226–27
worms
Archipheretima middletoni (earthworm), *xxvi*, 14–15
Halicephalobus mephisto (devil's worm), 170–71
Peripatus solorzanoi (velvet worm), *72*, 82–83
"Planocerid species 1" (flatworm), *144*, 164–65
Wunderpus photogenicus (octopus), *216*, 238–39

yams: *Dioscorea strydomiana*, *120*, 140–41
Yellowstone National Park, *168*, 176–77

Zambia: *Lepidiolamprologus mimicus* (fish), *192*, 204–5
zombie ant fungus (*Ophiocordyceps camponoti-rufipedis*), *144*, 154–55